To Ca
Good gym buddy
All the best

A Cowboy's Life

**MEMORIES OF A
WESTERN COWBOY IN
AN EMPIRE OF GRASS**

Happy Trails
Mack

by Mack Bryson

Copyright © 2013 by Mack Bryson
First Edition – January 2013

ISBN
978-1-77097-434-0 (Hardcover)
978-1-77097-435-7 (Paperback)
978-1-77097-436-4 (eBook)

All rights reserved.

No part of this publication may be reproduced in any form, or by any means, electronic or mechanical, including photocopying, recording, or any information browsing, storage, or retrieval system, without permission in writing from the publisher.

Produced by:

FriesenPress
Suite 300 – 852 Fort Street
Victoria, BC, Canada V8W 1H8

www.friesenpress.com

Distributed to the trade by The Ingram Book Company

Table of Contents

Bryson Family Tree vii

Carson/Magee Family Tree viii

Chapter One: My Home by the Fraser 1

Chapter Two: The Making of a Ranch 13

Chapter Three: Running a Ranch 29

Chapter Four: Our First (and Last) Beef Drive 41

Chapter Five: The Cowboy Takes a Wife
(or Mack Got Married & Lost the Cows) 50

Chapter Six: Shepherds & Thanksgiving 72

Chapter Seven: Things Not to Do with Your
Bride to Be & Things Not to do : with Your Wife 79

Chapter Eight: Beans Up–Beans Down 103

Chapter Nine: The Gold Bridge Bar 110

Chapter Ten: Dogs That Have Trusted Me 117

Chapter Eleven: Bull Shot 128

Chapter Twelve: Horses I Have Known and Respected 134

Chapter Thirteen: Falls I Have Taken 145

Chapter Fourteen: Big Bucks 156

Chapter Fifteen: Bear Bait 178

Chapter Sixteen: Incidents 183

Chapter Seventeen: Clarence Bryson:
Larger than Life and Revered : by His Family & Friends 196

Chapter Eighteen: Range War—Friend or Enemy 229

Chapter Nineteen: Jerry Eppler—
United States Marine 234

Chapter Twenty: CBC Tracks the Cowboys 237

Chapter Twenty-One: Empire of Grass 245

Chapter Twenty-Two: French Bar Canyon Rapids 252

Chapter Twenty-Three: Empire Valley Ranch—
A Huge Cast of Characters 263

Epilogue: Life after Empire 278

About the Author 292

Acknowledgements 294

Dedicated to my mother, Eleanore Bryson and my father Clarence Bryson, for raising me to be a good rancher and competent cowboy.

They taught me that your word is your bond and friends are more important than money.

Dedicated to my wife Elizabeth, whose love and patience have seen me over so many of life's hurdles. Without that patience this book would never have seen the light of day.

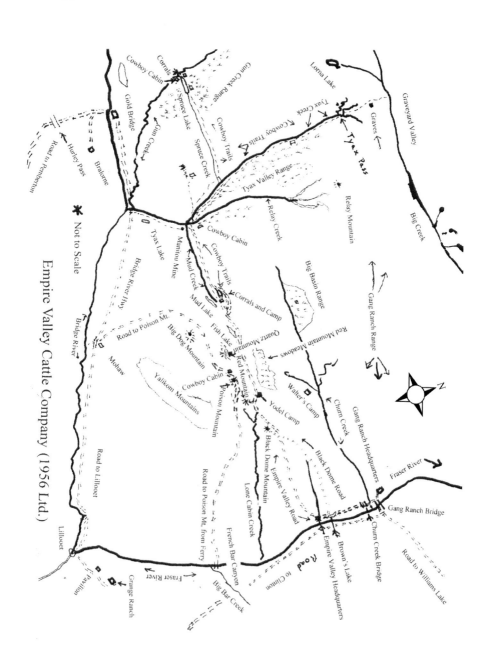

Empire Valley Sketch Map

Bryson Family Tree

Adam Bryson + Jane Stewart (came from Co. Antrim, Ireland to Nova Scotia in 1825) Their son:

John Stewart Bryson (b. Nova Scotia, 1827) **+ Elizabeth Catherine Bates** (b.1829 Nova Scotia) Their son:

John Bates "J.B." Bryson (b.1864, Nova Scotia) **+ Minnie Carson** (b. 1879, Pavilion B.C.) **married 1904**

J.B. Bryson had seven children:

Constance (by first wife, Mary Etta Currie) and with second wife Minnie Carson he had: Robert (Robin), Norval, Wilfrid (Duffy), Glen, Norma, & **Clarence Bryson** (b. 1908 Pavilion, B.C.) + **Eleanore McCallum** (b. 1910 Vancouver, B.C.). They married in 1931 and had three children: Donna (Bryson) Gillis, Duncan Bryson & Malcolm Bates (Mack) Bryson (b. 1933 Vancouver, B.C.) + Elizabeth Anne Stevenson (b. 1936 Toronto, Ontario) married 1960. They have three children: Lisa Jean (1961), Malcolm John (Jack) (1964) & Douglas Mackenzie (1966). They also have five grandchildren: Malcolm Eric, Duncan Mark, & Ingrid Elizabeth Bryson; and Malcolm Douglas Mackenzie (Mackie) and Aza Mairi Elisabeth Bryson-Bucci.

Carson/Magee Family Tree

Hugh Magee (b. 1822, Co. Fermanagh, Ireland**) + Isabella Crawford (**(b.1825, Fermanagh, Ireland). They were early pioneers in Vancouver.

Eliza Jane Magee (b. 1853, Ontario**) + Robert Carson (**b. 1841, Edinburgh, Scotland). They married in 1878 and were pioneers in Pavilion, BC, (in the Lillooet area) and had nine children:

William George, Robert Henry, Ernest Crawford, Frederick John, Helen (Ella), Eliza Jane (Girlie), Edith, Edna Ruth, and Minnie Carson (b. 1879, Pavilion, B.C.) **+ John Bates "JB" Bryson** (b.1864) married 1904

Chapter One

My Home by the Fraser

"Either the whiskey goes or I go."

These words were spoken by my newlywed great grandmother, Eliza Jane, upon her arrival at Robert Carson's ranch atop Pavilion Mountain Plateau in 1878. "Carson's Kingdom" it was called by Bruce Hutchison, in his epic chronicle *The Fraser*. A kingdom it was—although eventually the plateau was shared equally with my paternal grandfather, John Bates Bryson who married Minnie Carson the daughter of Robert and Eliza Jane. How all this came about makes for an interesting glimpse into BC history.

One story, as told to me by my father Clarence, has it that Great Grandfather Robert Carson had come west via the Oregon Trail with his family around 1855. They made a historic decision to travel from the East Coast of the US by wagon train to the West Coast, after emigrating from Scotland. On one of their stops, the story has it that as a young lad of fifteen, Robert was sent out hunting. When he returned with fresh meat, he discovered that most of his party had been slaughtered by a tribe of Indians- perhaps the Nez Perce tribe. Robert, it was said, caught on with the next wagon train and eventually arrived in Oregon, and from there he made his way north to the colony of British Columbia.

A second story tells that, while some wagon trains were meeting with a violent end on the Oregon Trail, Robert Carson's party (which did not include his family because he had run away from home at about age nineteen), evaded harm. He had been forbidden to leave for the west and so he decided to go on his own. His party was guided by an experienced wagon train master who tricked the Indians into believing they were encamped in one spot until dark. When darkness fell he would stealthily move his party further along the trail. In this manner he evaded the downfall of a party in front of them and behind them led by less experienced guides.

At left, riding without shoes, is Cataline- BC's famous pack train guide.

This story, I believe, is the true story, but I prefer the wagon train ambush "version" because it was told to me by my dad. Robert eventually arrived in the Colony of British Columbia in time to catch the "Gold Fever" which was luring the adventurous to the Northern BC goldfields — The Cariboo Gold Rush.

He crossed over from the States near Rock Creek, in the southern Okanagan, and got a job with a pack train operator. With this job he eventually arrived in Lillooet and started his own horse packing outfit to transport goods to the Cariboo goldfields from Lillooet, which was Mile 0 on the route to the goldfields. He would make his way up the Fraser River and from there up over Pavilion Mountain and down the other side to Clinton, which was Mile 57 on the so-called Gold Highway. At that point, he travelled north to Barkerville on the Cariboo Wagon Road, which later became the Cariboo Highway.

After several trips packing to the goldfields at Barkerville, he fell in love with the beauty and ranching potential of the Pavilion plateau, so in 1863 he decided to homestead there and develop the cattle ranching potential he saw in the region. In 1867 he was able to pre-empt land on Pavilion Mountain, and a few years later, after he had built up a herd of cattle, he and an old Indian cowboy and guide named Pacullah Kostah and a neighbouring cattle rancher named Richard Hoey made the first and only cattle drive of its kind to the coast, trailing 200 head of cattle down the hazardous Lillooet Trail, through the Capilano Valley watershed, to a terminus on Burrard Inlet near the present day Second Narrows Bridge.

The trail had been built by government contractors at the request of local ranchers, and roughly followed over a route where the railroad runs today. The route was extremely important for ranchers for they had no outlet for selling their cattle. Evidently the route the trail took was a political and environmental disaster perpetrated on the ranchers by an MLA—The Honourable Basil Humphreys. There was a rumour at the time that he had appropriated some of the $35,000 he had requisitioned from the BC government to build the trail. At times the cowboys had to pull themselves up steep hills on a very narrow rocky trail and the cattle had to climb up steps built by the government contractors, to get to the tops of those steep hills. When they finally reached Burrard Inlet, across from the city of Vancouver—then known as Gastown—several of their cows had died and most of the rest were too gaunt and thin to butcher because there was so little fodder on the arduous route. Gastown was the largest town west of the Mississippi and home to an obviously beef hungry populace. At the shore of the inlet, they built a scow barge to ferry the cattle across the inlet to the south, or Gastown side, where meat hungry Gastowners awaited them. They were able to sell some of the cattle that were fat enough to butcher. Those that were market ready were sold to a butcher named George Black. Coincidentally, my wife and I are now living in Vancouver on Wall Street, near the Second Narrows Bridge, overlooking the site of the end of the cattle drive.

Carson's cattle drive proved that the proposed route from the interior to the coast via the Lillooet Trail was not commercially feasible—rather it was a disaster. Luckily for the ranchers who were hoping to follow in the footsteps of that first cattle drive, the railroad arrived in Ashcroft in 1885, so the trail became redundant.

Luckily after he arrived at the coast, Robert found a farm which grew hay for wintering and fattening the majority of his cattle herd. The site of the Fraser River Farm was in the present day district of Kerrisdale in Vancouver. It was owned by a chap named Fitzgerald McCleery whose neighbouring farmers were Hugh and Isabella Magee. The Magees were among the first European pioneers in the Vancouver area. Hugh Magee, by some reports, was a bit of a cantankerous trouble maker, and is remembered as one of the colourful characters in the early days of Vancouver. Part of the Magee family farm is now the site of Magee High School in South Vancouver. Part of Robert's luck in finding Fraser River Farm was that his trips from Gastown to check on his cattle enabled him to visit with the Magee's eldest daughter, Eliza Jane. These trips became more frequent over the winter until the cattle were sold in the spring. He received a nice profit for his now-fattened beef and decided to return home with the promise to return the following year.

Robert headed back to his ranch, taking an easier, but longer, route via a new road recently built by the British Royal Corps of Engineers through the Fraser Valley. However, after a three day ride, he found himself thinking of Eliza Jane. He had ridden as far as Hope before he decided to return the one hundred miles back to Vancouver to ask Eliza Jane Magee to be his wife. One can only imagine her joy as she greeted her beloved, whom she expected to be gone for a year. She agreed to his proposal, and they were married at her home on April 3rd, 1878 by a Gastown Minister nicknamed "Old Hoisting Gear." Thus began a British Columbia pioneer ranching family dynasty.

The Carsons' honeymoon included days of travel by boat to Yale and then several harrowing days of travel in a stagecoach up the precipitous, dusty Cariboo Road through the Fraser Canyon. Much of the road, built by the Royal Engineers, was constructed by hanging wooden bridges around the rocky bluffs overhanging the mighty Fraser River. This route brought Eliza Jane to Pavilion Mountain for the first time on the stagecoach, where Robert was able to show her the ranch that he had built.

His home also served as a stopping house for stagecoach passengers bound for the Cariboo. The newlywed couple arrived at Carson's ranch house along with the other stage passengers to find a welcoming dinner catered by the ranch crew. Eliza Jane went upstairs to freshen up. When she returned in her finery, she was horrified to discover that the welcoming party had placed bottles of whiskey on the dinner table. Hitching up her bustle and putting on her coat she prepared to leave with her fellow passengers. Robert was taken by surprise. When he asked her where she was going, she pointed at the whiskey bottles and uttered those famous words, "Either the whiskey goes or I go."

Robert promptly took the whiskey bottles outside and smashed them on a rock. The rest is history. That was the last time liquor was served on the Carson Ranch. One would be right to assume that it took a very strong woman to make that challenge after such an exhausting trip. She passed on that mental strength and fortitude to her daughters and sons. And Robert, being an independent and strong-minded soul, was both a good match for Eliza Jane and a smart man to heed her request.

Eliza Jane put that strength to good use in many other ways. For example, in her later life when she moved back to Vancouver after the death of her husband, she was one of the instigators behind the construction of the Burrard Street Bridge. She was also a force behind instituting child labour laws and, not surprisingly, the Women's Christian Temperance Union.

Robert and Eliza Jane's eldest son, George, started the still-operating Carson Truck Lines. Unfortunately, he was killed as a young man when a truck motor he was working on fell on him as he worked on the engine from underneath.

Their second son, Robert Henry, was a Kamloops realtor who was elected many terms to the Legislature of BC in the Coalition Government of Robert Anscomb (Conservative) and Byron

"Boss" Johnson (Liberal). As a Liberal, Robert was elected to be the Speaker of the BC Legislature.

The youngest son, Ernest, owned the Pavilion General Store. This was the oldest operating general store in BC until it burned down in the year 2000, long after he had sold it. Ernest was elected to the Legislature of British Columbia as a Conservative from the riding of Lillooet, in the same Coalition Government as his brother Robert. Ernest is best remembered as the Minister of Public Works in the government responsible for constructing the Hope–Princeton Highway and the John Hart Highway from Prince George to Dawson Creek.

This was Ernie Carson's store and lodge shared with his wife, Halcyon and boys, Jack and Pat (my chum), and daughter, Dolly. Courtesy Don Carson.

Robert and Eliza Jane's oldest daughter, Minnie (born 1879) was my grandmother and married my grandfather, John Bates Bryson (born 1864) in 1904. Showing her mother's strength, when she was hospitalized for a serious heart attack she insisted that she would be just fine if the nurses would treat her with a mustard plaster over her heart. The nurses acceded to her request, knowing that a mustard plaster would not help a heart attack, but they were trying to make my grandmother comfortable. The mustard plaster did not work, and she died in 1941 at the age of 62.

My paternal grandfather, John Bates Bryson, referred to as Bates or JB, for his part was the third generation of a family who arrived in Nova Scotia from Northern Ireland in about 1825. My father like to joke that his family had lived on the English/Scottish border, where they caused so much havoc raiding for goods and food that the English soldiers scooped them up and transported them to Northern Ireland, where they became part of the so-called "Plantation Scots."

JB's grandfather Adam and his grandmother, Jane Stewart—along with Adam's three brothers and a sister—boarded ships bound for Nova Scotia around 1825. They arrived at the mouth of the Musquodoboit River and made their way up river to a promising looking farming spot called Middle Musquodoboit. Their farm was called Twin Oaks because of a home they built between oak trees. Those trees were still standing when I visited the house in 1997. To my eyes the farm looked small indeed to have sustained a large family. I had no trouble understanding why it was that my grandfather John Bates Bryson, his widowed mother, his brothers Melville, James, Allen, and his sister Ida came west to British Columbia.

Forebearers: Minnie and JB Bryson with four of their children. My father Clarence is standing on the far right.

John Bates started a Blacksmith shop in New Westminster, the new capital of the colony. There, in partnership with others of his persuasion, he started the first Liberal newspaper in BC. As an aside, when talking with my friends in Kamloops, Sandy McCurrach and Dean McLean, they told me that their grandfathers were also involved with that paper. Their grandfathers, also ardent Liberals, were living in New Westminster at the same time as my grandfather. It was not surprising then, that their grandsons, Sandy, Dean and I are still ardent Liberals and that we have worked together over many political campaigns.

It was in New Westminster that JB married a young woman named Mary Etta Currie, with whom he had fallen in love at first sight. They were living there very happily with their young daughter, Connie, when Mary Etta became ill and tragically died.

After his wife's death, a heartbroken JB got the roaming bug again and travelled to Ashcroft, leaving Connie with her mother's family. There he went into a blacksmith partnership with a fellow named Smith. Smith and Bryson shod the unbroken horses arriving from the prairies via the brand new Canadian Pacific Railroad. These semi-wild draft horses, after breaking and shoeing, were destined to pull 10-12 horse wagon trains up the Cariboo Wagon Road to Barkerville. With their earnings, John Bates and his partner bought a ranch called the Grange Ranch at Pavilion from a man named Captain Martley. Martley, according to legal documents of the times, was often in the Lillooet courthouse defending himself from allegations that he was taking too much water from Pavilion Creek—that water had to be shared with farmer licensees further along the creek. Ida Bryson reported that Captain Martley was the biggest man she had ever met. Robert Carson was one of those making the allegations which dragged Martley into the courthouse in Lillooet.

After they had operated the ranch for a year, Smith and Bryson split the partnership. Smith took over the blacksmith shop in Ashcroft, and JB took the Grange Ranch at Pavilion. Later JB added a Pavilion Mountain portion of the ranch which he purchased from a man named Clark. The close proximity of the Carson and Bryson ranches introduced JB to Robert and Eliza Jane Carson's daughter, Minnie, whom he courted and married.

Owing to the influence of Minnie and Eliza Jane, the Bryson family would be teetotallers for many years. My father, Clarence Bryson, the third son of that union, never took a drink until late in life. Following the family tradition, I did not have my first drink until I was 21 when my friend Roy Josephson phoned me from the Grand Hotel barroom in Merritt—where we ranched at that time—to tell me my Merritt friends were waiting to celebrate my birthday. I reluctantly joined the revellers, but for the several years following I was called "one bottle cap Bryson" for my reluctance to touch alcohol. Some would say I overcame that reluctance later while attending university.

JB also purchased a ranch in the Cariboo where he ran sheep in partnership with another of the Magee family. That ranch later formed a portion of the famous Empire Valley Ranch, which my father Clarence and our family purchased in its entirety in 1956 from the pioneer Koster family. Those sheep pastures were still referred to as the "Bryson Pasture" and "the Magee Flats" when we took over Empire Valley Ranch.

It certainly was a long and perilous journey from the Bryson-Carson beginnings in the old country to a new world here in British Columbia. The title of this chapter, apart from referring to the location of two out of the three ranches we owned, was the name of a song written by Keray Regan in 1948. *My Home by the Fraser* replaced even Bing Crosby's *White Christmas* at the top of the charts in BC. A very good local western singer, whose name escapes me, would sing that song at rodeos along the Fraser River that I attended when my dad was the rodeo announcer.

> *"My home by the Fraser keeps calling to me*
> *To come back, to come back to where I long to be.*
> *No dreamland or heaven could ever be found*
> *Like my home by the Fraser on that far distant shore.*
> *So far and yet so free, no other place for me,*
> *Than along the winding Fraser River shores."*

That is my favourite song, and I can still sing all the words to my friends. I don't know why, but they always say, "Don't give up your day job".

Chapter Two
The Making of a Ranch

My family purchased a very real ranching empire when we purchased the Empire Valley Ranch from Henry Koster in 1956. That ranching empire was comprised of 30,000 acres of prime ranch land. The range fronted on the Fraser River for twenty five miles from Churn Creek to Lone Cabin Creek. It was primarily covered with blue bunch wheatgrass, needle-and-thread grass, and fescues, with the upper, timbered parts covered with Kentucky bluegrass, June grass, and other various fescues.

Boyle Ranch. Empire headquarters is just out of sight at the end of the fields

The portion of the ranch that the government (hence the people of BC) owned was referred to as Permit Land, meaning we ranchers were permitted to run a certain number of cows per day at a nominal amount of rent per month. As I recall, it was about 75 cents per cow per month in 1956 and eventually ran around $2.50 to $3.00 per cow per month in 1998. Our Permit Land at that time consisted mainly of Big Basin, parts of Little Churn Creek, the New Pasture (now called the Maytag Pasture), the China Lake Pasture (now called Koster Lake), Fareless Creek Meadows, Black Dome Mountain, Starvation Canyon, and, most importantly, Red Mountain Meadows.

This last portion could graze 250 head of cows in summer—that was if Henry Koster's contractor, a man named Alfie Higginbottom, could keep the swamp grass meadows clear of a weed called larkspur. Alfie and his boys would pull up the weed by hand. The invitingly beautiful purple flowered weed was poisonous to cattle and so it was a laborious but very necessary job. Alfie did a lot of contracting for Henry and in fall he guided hunters out of a little cabin just beyond the ranch property's perimeter fence.

All told, the ranch was composed of around 88,000 acres of Crown Granted land (land that we owned) and Permit land in 1957. This healthy total was actually about 10,000 acres of lodge pole pine (commonly, but erroneously, called jackpine) and swamp grass meadows plus the 30,000 acres that comprised the home ranch—the rest was pretty good grass-covered cow pastures. (Jack pines only exist in the north west corner of BC.)

So it was that, as our herd expanded, there was just not enough grazing land for the cattle. We were faced with either finding new range or finding some method of intensifying our hay fields. Since there was ample evidence to indicate new ranges beckoning, or "Grass Beyond the Mountains" to quote from Rich Hobson's book of that name, the former option seemed the most alluring in 1958.

By 1958 my brother-in-law Don Gillis and my sister Donna had joined us as partners on the Empire Ranch. They had previously been running the Glen Walker Ranch on the Coldwater River near Merritt for Don's father, Dr. J.J. Gillis. When Dr. JJ decided he would sell the ranch Don, Donna, and children joined us as partners on the Empire Ranch.

Walter Grinder was the son of a long time Big Bar Creek rancher and his First Nations wife. Walter had arrived at Empire not long after we took over, riding in on his big brown horse, Billy Boy, looking for a job. He became one of our best long term employees on the ranch, and a good friend.

With a little preliminary planning, Don plus Walter Grinder and I embarked on a new range-finding journey which eventually netted Empire Valley Ranch half a million acres of some of the wildest and most beautiful backcountry Government Permit Range available. To put the area in perspective, that is almost the size of the Fraser Valley from Hope to Surrey. The process of acquiring, managing, and servicing that huge country forms the basis of many of my stories to follow.

That summer of 1958 the three of us followed a Forest Service map through some of the most rugged parts of BC and found some of the most beautiful rangeland in the province. Embarking from Red Mountain Meadows at the western extremity of the ranch that had been put together by the former owner Henry Koster, we horse-hoofed it west to the upper reaches of Churn Creek, through the Fish Lake country, to Mud Lake, and from there down Mud Creek to a point where it joins Relay Creek and the Tyax River. This is where past miners had established a long abandoned mine called the Manitou Mine. At this point, three small waterways become the Tyax River, which then jostles its way south through immense boulders and the resulting rapids to a confluence with the Bridge River near Gold Bridge.

We camped one night at the Manitou Mine and then followed an old mine road to a ford across Relay Creek. From here, climbing into the Tyax Valley on a near vertical trail was very demanding. However, as it later turned out, once my father had managed to build a switchback road up the hill, we could actually use that route to move cattle up into the Tyax Valley from the Relay Creek plain.

Eventually we arrived at a point on our map where the Tyax met a small tributary called Spruce Creek. We followed a trail up a brushy, meadow-bordered creek to reach Spruce Lake. In those days, Spruce Lake was practically unknown, except to a few packers and sportsmen. The most notable of these was a packer and wildlife photographer called Charlie Cunningham, whose pictures of the area at the time are quite famous. In my estimation, the best of these in the Lillooet Museum is an autumn shot showing forty to fifty Bull Moose single filing off the Dill Dill Plateau in the South Chilcotin Mountains, to start their annual rut.

On the edge of Spruce Lake was a cabin built by a famous guide named WG (Big Bill) Davidson from old, dry standing deadfalls. It was at this old cabin that we stopped for lunch. It had a magnificent view through the one window to view the lake with majestic snow covered Mount Dixon in the background. After lunch Don and I ventured down to the lake shore and encountered a leaky old rowboat. As lots of fish were jumping, and as we had seen rods in the cabin, we quickly planned an impromptu fishing expedition with greedy thoughts of fat trout for dinner. Walter told us he would have nothing to do with this, for to him it seemed a perilous venture.

Davidson Cabin on Spruce Lake with cattle drive headed to the Gun Creek range.

We poled our leaky launch down the lake with a pine bough to discover another unused and badly deteriorating cabin a half mile down the lake. Since no fish were biting our bait, we ventured ashore. Like Davidson's old cabin, this one was also unlocked, so Don and I entered into the dingy, rat-smelling interior. The dusty, cobweb-covered windows were draped from the inside with burlap sacking to deter bears from guessing what goodies might be cached inside. About the time that we had assured ourselves that the cabin could not be redeemed for a future cow camp, we could hear Walter Grinder leading our pack string up to the cabin as he searched for his cowboy fishermen friends. No doubt a superstitious Walter was not anxious to be by himself in this strange, though extremely beautiful country. No doubt he probably felt that we had drowned in our leaky old wooden rowboat. He himself couldn't swim and would never have entertained thoughts of venturing so far out onto the lake.

As soon as I heard Walter coming I came up with the idea to try to spook him. Quickly we flipped the burlap sacking back into place, leaving one window uncovered to allow Walter to peer in. Then we scampered up the ladder into a sleeping loft. When Walter peered in the window, shading his eyes to acclimatize them for viewing the murky interior, I issued forth a low groan, which to Walter must have sounded like a dying moose, or worse yet, the spirit of the former cabin resident. Walter snapped away from the window but stuck his head back for one last, anxiety ridden peek. Again I gave out with my rendition of a departed resident's spirit and just as suddenly, Walter was gone.

By the time we got down from the loft and out the door, Walter was half way across the clearing, dragging his pack horses and our saddle horses behind him at the gallop, while whipping his horse, Billy Boy, across both sides with his halter shank. Never had I seen a dark-skinned Native go completely white in the space of a few short moments. Later I felt a bit guilty about the charade but the moment remains vividly etched in my memory. And if the ghost of Walter, who was tragically killed in a traffic accident some years later in Williams Lake, is looking down on me now, I ask your forgiveness, Walter, but I still have to chuckle every time I think about that terrified look on your face.

Ghost cabin—with Walter, Don, and Francis at first cowboy tent camp

We did finally entice Walter to return to the cabin. Some very loud hollering from me finally managed to get his attention to alert him that we were in fact alive and not apparitions after all. This was lucky for us because he had both our saddle horses, all the food and camp gear we owned and was headed down the long rocky trail to Gold Bridge at a gallop to report his buddies lost to the spirit world. Walking that country in cowboy boots is no fun at the best of times, and it would have been a painful twenty mile hike for us from Spruce Lake to Gold Bridge down that rough Gun Creek trail, though it would have served us right for playing such a bad trick on a good friend.

Since our travels had taken us through enough good grazing land to consider applying to the Grazing Division of the BC Forest Service to run cattle on this route, we decided to head home. We branched off up the Taseko River to its headwaters and then over the Tyax Pass to descend into the Graveyard

Valley. This is a beautiful part of BC and an integral part of the famous Gang Ranch. Here, after some searching, we found the ancient mounds covered with decaying logs that had given the valley its name. As I understand it, in the old days the valley was part of the Grease Trail where the Chilcotin Indians packed their ochre paint down the trail to trade for the greasy oolichan fish on the coast of BC. Story has it that one band picked up the smallpox virus which was then raging on the West Coast of the Colony of BC and the Colony of Vancouver Island. On the way back with the prized oolichan grease the band was overcome by smallpox. Consequently many of the band members perished and were buried here in what is now called Graveyard Valley. We carried on down Big Creek till we turned off on a trail to Big Basin which was Empire range and then crossed Churn Creek for a three hour ride to home. I immediately contacted Alf Bawtree of the BC Forest Service in Kamloops who later accompanied me out to inspect these new ranges. At that point he acquiesced to our request to be granted these new ranges. Later, we also acquired the majority of the old Hayward Sheep Range.

About this time Francis Haller, like Walter Grinder before him, rode up river from Big Bar looking for a job. Francis had been dad's head cowboy when he managed the Diamond S ranch at Pavilion. He was immediately hired as our head cowboy. Francis would look after the herds of cattle that trailed out to the back country from Empire

Jumping ahead to a few years later, that old "ghost ridden" cabin site on Spruce Lake was the scene for another spooky happening. After securing a Forestry range permit to run our cattle in the area, we had moved a herd of heifers and young bulls from Empire to what we called the Gun Creek Range, and we were camped on the Davidson property at the other end of Spruce Lake. On this particular trip, Walter Grinder had brought his nephew from the Chilcotin with him to help him in his wrangler duties. The fourteen year old boy was hired to help Walter wrangle the spare horses and pack string. Francis Haller, to his great delight, had used the occasion to frighten the poor young fellow almost out of his mind with stories of the "hairy women" who were always looking to make off with young boys.

Before heading back to the ranch to bring out another herd, we decided to investigate the Little El Dorado Creek watershed, which comes out of the El Dorado Mountains, west of Spruce Lake. We left the boy behind to watch the camp and pack firewood. After discovering and wandering through some of the more majestic and beautiful parts of BC, we returned to find the camp empty. Fearing the worst, we began calling for the boy and were finally rewarded by hearing him answer from the uppermost branches of a nearby spruce tree where the beleaguered lad had climbed to escape any marauding hairy women who were reputed to be unable to climb spruce trees.

Spruce Lake and the Gun Creek area were, and are, some of the most beautiful parts of our province, so we were always happy to arrive there each year, late in June, with about 450 head of heifers and 20 bulls. Usually Francis Haller was left at the cowboy cabin on Gun Creek, at the foot of the Spruce lake sidehills, to manage the cattle in the area. I had managed to come up with five hundred dollars to pay a friend, Joe Bingham, for the Gun Creek cabin, which he owned but did not need for his guiding operation in the Spruce Lake/Tyax region. Joe's headquarters was an old abandoned resort on Tyax Lake, also called Tyaughton Lake. (Now, after Joe has passed on, his old lodge has become a high end resort, Tyax Lake Lodge.)

Looking after our cattle was a big job for Francis. It entailed constantly moving bunches of cattle to find fresh grass and preventing them from all congregating around Spruce Lake itself so that we could preserve some pasture for the horses of packers and guides in the fall. Francis also had to keep a close eye on predators, both two legged and four legged, for at Spruce Lake we were only about twenty miles from Gold Bridge, and sometimes unemployed, meat-hungry miners would ride in and lick their chops while looking at those fat lazy heifers. I am sure that more than one entertained thoughts of beefsteak broiled over a bed of campfire coals. That is, until they found Francis dogging their tracks with his .30-30 carbine slung from his saddle.

Another of Francis's chores was to break up the young bulls from getting together and mounting one another instead of the virgin heifers. Most people are unaware of the fact that young bulls are without gender preference, and we could not afford to let our female stock go unbred. Again, jumping ahead a few years—the Gun Creek Range with its lush grasses could turn out some wonderfully fat beef by fall, but these same animals, after they had been driven the 90 rough miles back to the home ranch would not always arrive in peak market condition. At that time we normally pregnancy tested the heifers at Empire Valley and shipped the "empties" by truck to Ashcroft where they were loaded on railroad cars for shipment to feed lots in Alberta. It eventually occurred to me that, if we built corrals near our cowboy cabin on Gun Creek, I might get my friendly veterinarian, Lorne Greenaway (later to be elected the Progressive Conservative Member of Parliament for Coast Chilcotin) to come to Spruce Lake in late August to pregnancy test the heifers, under the guise of a fishing trip.

My phone call, as it turned out, came at just the right time for Lorne had recently purchased a very expensive fly fishing rod which he was anxious to try out. Thoughts of fly fishing for the famous Spruce Lake trout immediately captured his interest. If he would come, we could then herd the non-pregnant (empty) heifers down the Gun Creek trail to the corrals and loading chutes on the Gold Bridge Highway that had been built by BC Electric—the forerunner of B.C.Hydro—to truck their cattle from their base at Lillooet to graze them on the Gun Creek range. Lorne acquiesced to this plan, so the cowboys and I spent two weeks building a large set of corrals complete with catch pen and chute near Spruce Lake on the Gun Creek itself.

Don Gillis crossing cattle on the Tyax River at low water

Everything was made from materials found at the site. We chain sawed the longest, straightest young spruce trees we could find and dragged them behind two cowboys on horseback. We peeled the logs and, with great effort, devised a system of rolling the logs up a ramp to be placed one at a time on top of one another in a zigzag fashion up to a final height of about six feet. The chute itself was built with posts dug into the ground with a space for the veterinarian to walk into the chute behind a heifer that had been contained and constrained by a pole squeeze. The pole squeeze acted like large constraining scissors. The scissors action was accomplished through the use of two sturdy fourteen-foot poles joined together at the base with nylon rope and reaching straight up so that a man could stand upright up above the cattle on a pole platform where he could operate the tops of the squeeze poles, one with each hand. He had then only to wait for a driven heifer to put her head through the chute exit, whereupon, he could quickly close the two poles together around her neck. A bit of safety rope could then be secured around both poles at chest height, and the cow or heifer, whichever it may be, was effectively restrained for examination by the veterinarian.

After the corral was built, I rode down to Gold Bridge and phoned Lorne to come in and do the tests, repeating my promise of some great trout fishing on Spruce Lake for his troubles. Lorne was anxious to come and arrived soon thereafter with his beautiful new fly rod. Unfortunately, when the rod was broken down into its component parts, it still stuck out from under the pack of the horse that carried his gear. Both Lorne and I were aggrieved to learn that the forward end of his rod case snagged on a stout jack pine bough and shattered his prized new fly rod into several pieces. That was an all too quick and unhappy ending to Lorne's promised fly fishing adventure. As things turned out, we ended up being too busy to do much fly fishing anyway.

On that particular drive I had brought in a friend with the crew from the home ranch to help us out. Jim Davies was then managing partner of the Chilcotin Inn at Williams Lake. Donald Gillis and I had met him first at "Willie's Puddle" and liked the big ex-football player as much as he loved the rough and tumble life of a cowboy. For entertainment he and I used to wrestle around the campfire to the delight and occasional accidental dusting of the cowboys. By the time Lorne and I arrived back, Jim and the rest of the cowboys had gathered together all 450 head of heifers and put them in our new corral, over which we were all busting our buttons with pride. Despite the sturdy look of our new project, we were surprised to find that our herd was a bit too rambunctious. Upon first entering the corral they surged to one side and forced a break in our new fence, which required rebuilding and strengthening before the stray cattle could be brought back into the repaired corral.

This time the fence held, and we could push cattle into the catch pen and from there down our chute with the pole squeeze at the egress. The whole thing worked like a damn. Lorne was able to step in behind each animal and do his pregnancy test without much difficulty. My job was to operate the poles and stand above the chute. As the animals were pushed down the chute by the cowboys, I would nail them around the neck with the squeeze. The job required good timing and a pretty strong pair of biceps to squeeze the poles together until I could wrap the safety rope around the poles and take the pressure off my arms.

After a few hundred females had been pregnancy tested in this manner, big Jim Davies volunteered to take over the tough job at the squeeze end. The truth was he was getting a little too well greased with fresh cow manure as he coaxed and prodded the animals up the chute by twisting their tails. I humoured Jim and gave up my job standing above the heifers in the fresh air. Jim had a little trouble at first, getting the hang of waiting until the heifer's head popped out of the front of the chute, and he lost a few head to "late squeezing." This required a rider to take off after, rope, and return stray animals back to the corral gate.

I gave Jim some coarse instruction on how to do the job properly if he was to continue in the position of power. This caused Jim to lean forward on the ramp and peer down so he could anticipate the next animal to enter the squeeze in time to slam the poles together. The only problem was that as he leaned forward, he held the poles on either side of his head with those brawny arms of his. The next heifer that jumped forward to escape hesitated at the final second in the chute but not before Jim, anticipating her escape, had slammed the poles together on either side of his ears. The impact knocked him unconscious. I can still recall watching Jim's hands sliding down the poles before he toppled helplessly to the ground. Jim had always loved to torment anyone who couldn't measure up to his high standard of manliness, so, after we ascertained that Jim was relatively unscathed, the whole crew and myself had to stop action for nearly ten minutes while we rolled around, laughing at the sight of an angry, groggy Jim. Truly, it couldn't have happened to a more deserving guy!

In the end we did drive the empties down the Gun Creek trail, after I had returned Lorne to his vehicle. I phoned Baldy Boyd Trucking, based in Clinton, to bring in two cattle trucks to the corrals on the Gold Bridge highway. We cut short (banged) the tails of the empties as well as those showing to be late calvers, for we had to turn the whole herd loose after testing and then round them up a few days later to separate the empties and late calvers for shipment.

The whole exercise was a lot of work, and we only did it just that one time. The last time I saw them, in 1999/2000, the corrals we built were rotting but still being employed by packers to corral their pack horses overnight. The cowboy cabin is also in very rough shape; the roof is falling in, and it's full of packrats. The Gun Creek range is no longer used by Empire Valley Ranch as the range was given over to the packers and game guides for their horse herds. The new owner of Empire Valley traded for some range south and east of the Manitou Mine in a trade that was considered a good one because it was closer to home and it had been logged over and re-grassed. In my estimation, no cattle range could ever replace the one on the Gun Creek sidehills where we grazed our cattle. However, I admit to being biased because of the beauty of that splendid country and the good times and the good friends we made at the Spruce Lake and Gun Creek cow camps.

Top of the Eldorado Mountains—horse packers with Tyax range behind

A footnote to this saga is that one of those years Francis Haller showed up at the ranch saying he wanted to hire a couple of young Gold Bridge men to "teach them the ropes" and help with the herd. I acquiesced and that introduced us to Barry Menhinnick, who in the year 2000 had his own packing and guiding business out of Gold Bridge which he calls Menhinnick

Mountain Rides, Chilcotin Wilderness Adventures and Spruce Lake Trails. He operates this with his son Warren. Barry was a quick study who worked for a couple years with us until the ranch was sold in 1967. It's kind of strange for me now, when I go back up to that country, to see and hear Barry refer to himself as an old Mountain Man[1] and it causes me to wonder— *what then I have become, for I am many years his senior?*

—[1] Robert Bateman and Terry Jacks made a video to promote the establishment of a provincial park surrounding Spruce Lake/Gun Creek/Tyax. The video shows a whiskered Barry Menhinnick holding Terry's young son while he tells the child a story about a famous Grizzly Bear, wherein he tells the boy to listen closely to the "Old Mountain Man".

Dunc, Dad (Clarence) & Mack in front yard of the Home Ranch at Empire Valley, 1957

Chapter Three
Running a Ranch

In a sense, the empire that the Bryson family took over in 1956 could be classified as an "old" empire, for the ongoing operation was rooted in BC's early ranching day methods. The previous owner, Henry Koster, had teams of Clydesdale horses that pulled hay wagons, horse mowers, dump rakes, stone boats, and so on. Henry had upgraded some by purchasing a D-2 Cat and an old tractor which pulled a sometimes-functional hay baler, but most of the ranch work was done with horses. Henry's favourite purchase had been a Piper Super Cub airplane, which he and I took out for a spin over the ranch lands almost every evening when I first arrived at the ranch.

Unfortunately for the Brysons, the price of cattle plunged the year we bought the ranch, and in addition to our down-payment of $75,000, we had to make an initial $75,000 mortgage payment to Henry that fall. We sold off every spare head of cattle on the ranch, except the basic cow herd of five hundred cows, but the returns still didn't tally up to what we needed, and so we had to have Henry take back the D-2 Cat and the Super Cub to complete the payment. I did so reluctantly, for I dearly wanted to learn to fly that Super Cub.

The Cat, which was an essential for the ranch was not replaced until my father in law, Doug Stevenson of Mackenzies Ltd, loaned us $6,500 to buy a renovated TD-9 tractor from Finning Tractor in Williams Lake. The manager of Finning Tractor was Bernie Moore and he and his wife Jean became long-time friends of ours. (As a matter of fact, years later, Bernie and Jean's daughter Megan married my cousin Layton Bryson, son of Duffy and Dorothy. Layton and Megan took over Duffy and Dorothy's ranch at West Pavilion.) Bernie went out of his way to make sure that our used TD-9 Cat was in the best possible mechanical condition. The amount of work Clarence did on the ranch with that TD-9 was staggering. It was operated almost every day, summer and winter.

With that TD-9 we built 65 miles of road onto the end of the existing 28 miles leading west of the ranch. That road became our lifeline to the backcountry for moving cattle and supplies, summer and fall, to our Tyax and Gun Creek ranges. Later on, the BC Forest Service joined our roads up with a network that allowed the public access from Big Bar across the Fraser River Ferry and also up the Mohaw Road from Lillooet. Not only that, they also allotted us, under their range improvement service, $1,800 to build a cow trail from Empire to the Tyax Valley. Instead of using men with axes to build the trail, we used the TD-9 to build a rough four-wheel drive road that Clarence upgraded every year with his Cat.

To build that road, my brother Dunc and I alternated weeks running a chainsaw in front of Dad on the TD-9. We started the road in early summer and completed the 65 miles of four-wheel drive road by fall. In addition to maintaining that road, Clarence also maintained the public road leading to our headquarters from Churn Creek. Though this road was also maintained by the Public Works Department from Clinton, they could not be counted on to fill in all the constantly occurring washouts in spring, or to plow out the constantly drifting snow in winter.

The job I hated most was also taken over by the Cat, namely the clearing of approximately 20 miles of irrigation ditches that serviced the ranch hayfields. By hand, it took a back breaking two or three weeks of work for a crew of four or five in spring. Clarence would clear and build new ditches all by himself in about a week with the TD-9; however, I doubt he had as many amusing incidents with that Cat as we did by hand.

One spring my friend Nick Kalyk came to the ranch to help get some good old fashioned ranching experience. Nick was a buddy from my UBC Faculty of Agriculture days, who was teaching agriculture, science, and math at Kamloops Secondary School in Kamloops. Nick and I were good buddies and we took every opportunity to test one another's physical prowess. After a couple of days hacking at rose bushes and shovelling out couch grass, Nick and I became bored and flew at one another in the ditch alongside the Boyle Ranch hayfields. We fought for two hours and mowed down a quarter mile of rose bushes. Needless to say brother Dunc, Don Gillis, and Walter Grinder were not amused because we also didn't clean the stretch of ditch assigned to us that day, and they had to finish it for us.

Grinder Creek Branding Corral—Nick on Spooks

On the ranch I prided myself on being a fairly good teacher about ranch life and so one day I was showing my nephew, John Gillis, how to catch a grouse. On this particular day I was driving a tractor with Don and his son John riding behind on the trailer. We were going to haul some hay and we were passing through a field of long grass and alfalfa. I noticed some grass waving as though a grouse or a pheasant was making its way in front of the tractor. I told John, "Watch this. This is how you catch a grouse in the grass." I pulled up alongside of the waving grass and stood up on the wheel of the tractor. I launched myself with arms outstretched and legs in a curve to wrap around a grouse. However, when I grasped the victim, it turned out to be a porcupine. It took Don half an hour, using pliers, to pull out all the porcupine quills from my hands and legs. I presume John learned a good lesson here.

Running the Cat alone was not something we encouraged after Dad had an unfortunate incident which nearly cost him his life.

He was building a bridge over a washout close to the main buildings and was using the two and a half inch cable attached to a winch on the TD-9 to tow timber beams across the washout. Having finished the job, he stepped off the Cat to recoil the winch line. He was standing behind the Cat, facing it in order to feed the cable onto the winch spool while the winch clutch was engaged. He had the powerful diesel motor idling, but the cable winch travelled fairly quickly, and he was trying to ensure that the cable didn't end up all on one side of the drum, or all tangled up. To keep the cable straight, he was leaning back and pulling on it with his gloved hands as he fed the cable onto the drum. A little slack in the line caused him to step back and pull out the slack.

When he stepped back, he stepped right into a coiled up bight in the line. Quick as a wink, the coil ran up and tightened around his upper leg. Dad was inches away from being wound around the winch drum when he blacked out from the pain. About that time the pressure of the winch on his body caused the cable to flip him upside down and out of the bight of the cable. He came to with a terribly bruised leg and thigh but never missed a day's work on the Cat. The old boy was pretty tough and set a standard of work that kept us all busy and tired but tough at the end of every fourteen hour working day.

I well remember haying on the Diamond S Ranch at Pavilion Mountain as a boy from 1944 to 1946 (ages twelve to fourteen). At that time we had a crew of twenty to thirty men and ten to fifteen teams of horses putting up 75 tons of hay a day. After a day or two of curing in the sun, we used horse sweeps to bunch the hay that had been left in windrows by me on the horse-pulled dump rake. The sweeps would pick up the bunched haycocks and deposit them on the teeth of an Overshot Stacker, where two or three men tried valiantly to form a bread loaf shaped stack before being buried by the next half ton of hay flying up at them. This was accomplished with the help of the giant flexing spring-loaded hand of the Overshot Stacker. When we were stacking hay, my job was to run the derrick horses out about twenty-five yards from the stack. Attached to the double tree behind the horses' harness was a cable running back through a system of pulleys that caused the giant wooden arms to raise and throw the hay onto the top of the stack. That system, of course, was light years ahead of the early 1900's system which required men to cock the hay from the windrows with pitchforks and then fork those onto slings attached to horse drawn sloops or sleighs. At the stack, the slings were then attached to the hook of a cable block and hoisted into the air by a team of derrick horses using a system called a Boom Stacker. We used that system before we got into using the Overshot Stackers.

My job with the Boom Stacker had been to run the derrick team out fifteen or twenty yards to lift the hay up under the boom of the stacker and over the heads of the men. I would trail a trip rope behind me and when the stackers hollered "Trip er!" I would jerk the rope, causing the slings to split apart in the bottom of the load. The hay would then drop onto the exact spot selected by the stackers. Occasionally I would step on the trip rope at an inopportune time and a thousand pounds of hay would fall upon some sweaty and unsuspecting stackers. Both this action and having my derrick team run away was met with a good chewing out by the workers even though I was the boss's son. The worst admonition I recall was "*Cultus shama!*" which means bad white man.

Anyway, back to the Empire Valley horses for a moment. The horses filled a need admirably but we had to get our operation into modern haying methods. It didn't take long until we were able to secure a $6,000 operating loan for the ranch from the Bank of Montreal in Clinton so that we could purchase our first Massey Ferguson Tractor. This tractor was still operating when we sold the ranch twelve years later. There is a really nice feeling about sitting on a horse mower seat listening to the snorting of the horses and the chattering of the hay knife moving across the serrated guards. However, comparing production of six to eight acres a day from a team of horses just doesn't measure up to the sixty to eighty acres a day of production from a tractor mower with a six foot blade.

With two tractors, a modern hay baler and a crew of two we could put up 50 tons of hay a day. The labour part of haying was made even easier when we were finally able to afford to buy an automatic hay stacker. I appreciated that as much as any piece of equipment on the ranch, for I had spent too many years throwing fifty to seventy five pound bales of hay over my head to build a stack of hay. So much for the old Empire's pleasant, but archaic way of haying—our family had to operate the ranch in a more modern way.

FROM THE OLD TO THE NEW— MANY DEALS LATER.

After we had sold Empire Valley to Maytag in 1967, the ranch was run pretty well by Bob Maytag's foreman, a man named Feldhauer, whose children I later taught in Agriculture at Valleyview Junior Secondary School in Kamloops. Bob Maytag held the ranch for four or five years until he felt he had lost enough money ranching and sold the ranch to a lady named Sophie Steggeman from Munich, Germany.

Steggeman and her son and daughter had grown up on a small farm on the outskirts of Munich, which became industrial land.

The fortunate sale of her farm to German industry enabled Sophie to purchase this very large operating cattle ranch in the BC Cariboo/Chilcotin country. She didn't have the first idea how to manage the ranch so she would invite successful rancher neighbours to advise her. My invitation followed visits by Henry Koster, the Sidwells who managed the huge Gang Ranch, and even Joe Gardner, the manager of the famous Douglas Lake Cattle Company near Merritt.

Sophie ignored everybody's advice, including mine. When I arrived at the ranch after driving in from William Lake, I was surprised to find Sophie and her two children cocking hay with pitchforks the way they had on their twenty acres outside of Munich. Needless to say, trying to cock hay on 600 acres of Empire Valley's alfalfa crop land was just not going to get the job done.

Sophie listened carefully to what I had to say, and then did just the opposite because she knew better. Contrary to traditional Cariboo hospitality, although I had spent the day with her, I was not offered lunch or dinner. Her own adversity caused her to forfeit the ranch back to Bob Maytag under a bankruptcy order a year later. The old gal wasn't stupid, though; she had put $400,000 down on the ranch, and as far as I know had never made a mortgage payment. Maytag's Kamloops lawyer hadn't included cattle or machinery as collateral on the mortgage, so when the bankruptcy forced her off, she took all the machinery, all the tools and approximately one thousand head of cattle and headed for Kamloops where she alienated many Thompson Riverfront hobby farmers as she pastured her cattle on any poorly fenced farmland along the river.

In my estimation she probably came out ahead on that deal by at least $100,000 while she took advantage of those landowners along the Thompson River east of Kamloops. Maytag, for his part then traded ranches with a guy named Hook from Colorado, who brought in a set of aluminum sprinkler pipes and put the whole of the Empire Ranch's hayfields under gravity flow sprinklers. I think he and his family enjoyed the ranch, but the loneliness and the lack of schooling for his children drove him to sell after taking off a reputed million dollars' worth of timber. Hook again traded ranch for ranch with a guy named Pepperling from Oregon.

Now we must go back a few years to get to the real meat of this story—we were able to purchase the ranch in 1956 because of a complicated deal. This deal involved selling several thousand acres of timberland which belonged not to us, but to Henry Koster's Empire Valley Ranch. This came about in the following fashion: Bob Carson, my dad's cousin, was a realtor in Kamloops who had listed the Empire Valley Ranch for sale from Henry Koster. Bob came to Dad while we were operating the old Voght Ranch in Merritt, where we owned 150 acres of hayfields and 3,500 acres of rangeland along what is now the Coquihalla Highway. Those same hay lands today contain a quarter of the City of Merritt including schools, apartment buildings, parks and such, and you could not now purchase a single lot there for the $35,000 we got for selling to Ken and Doris Gardner and sons Jerry, Jimmy and Ross. (Unfortunately for them they did not realize much of the big money either as they sold before the rich Craigmont Copper Mine opened nearby and that opening created a bonanza for the new owners.)

At the time, Bob Carson thought we could put a deal together on Empire Valley if we could come up with the $75,000 down payment required by Henry Koster. To realize this, Bob and Dad went to Vancouver to see if they could find someone willing to loan us some money.

Dad had run in the Provincial election of 1952, when WAC Bennett—Wacky Bennett to his detractors—had come to power over the prostrated Coalition bodies of the Conservatives under Herbert Anscomb and the Liberals' "Boss" Johnson. Dad's Provincial Liberal Leader in 1952, was Gordon "Bull of the Woods" Gibson, who was quite taken with the big flamboyant rancher. They were both defeated.

Gordon had promised Dad if he ever needed money or anything else, to come and see him. Gordon Gibson's large fortune had come from whaling in Hawaii and logging on the BC coast. Dad's idea was to see if Gordon would be interested in loaning the needed $75,000. Dad arranged a meeting with him in Vancouver. Gordon claimed he couldn't quite swing the necessary funds by himself, but he would pull in a couple of his rich buddies to make sure the necessary funds would be made available. Dad was asked to come back the next afternoon to have a meeting with them. At that time the offer presented to Dad was that these corporate big wigs would lend him the $75,000 for a 51% interest in the ranch. Dad was not about to give up financial control if we were to buy the ranch so he told them that he would have to phone and consult the family to see if this was satisfactory. He had no intention of phoning anyone and was so incensed by his old mentor's offer that he walked out and was fond of saying, "I hope they're still sitting there waiting for my answer."

Then, between Bob and himself, they got the bright idea of selling some timber for the down payment. As luck would have it, they chanced on the big coastal logging firm of MacMillan Bloedel, who were serious about getting away from the coast of BC to gain a foothold in the interior forests. MacMillan sent a Goose (an amphibious plane) and a team of foresters to Empire Valley. They landed on the natural 3,000 foot airstrip on Bishop Mountain alongside the potential timber sale. The foresters quickly got back by radio phone to their headquarters and reported that it would be a good deal. A cheque for $75,000 was soon handed over once the proper legal arrangements had been made. We got the Empire Valley Ranch and MacBlo got 10,000 acres of tree farm. Under the agreement the property could be repurchased by Empire Valley after the future logging was completed. The price to us for the buyback would be one dollar per acre, in perpetuity.

When the Bryson family sold Empire Valley to Bob Maytag in 1967 the repurchase agreement was not transferred over to the new company because of a lapse on the part of Maytag's lawyer—now a judge. When the error was discovered sometime later, the Maytag lawyers came to our lawyer in Kamloops. We were told that we had to sign over the lease with no recourse of any monetary value or they would sue us for lack of performance. I objected on the basis that any time that I had any dealings like that it required some money to change hands. Our lawyer felt we could be sued if we tried this, so we eventually signed off on the deal without additional consideration. The next year Maytag's lawyers were back again because they still hadn't registered the easement repurchase agreement. Again I protested that we should be compensated, and again our lawyer convinced us to just sign the paperwork.

Not too long after this my dad had a stroke and was no longer able to be president of our company. Later, Maytag sold the Empire Ranch to Hook, and a few years later Hook's Kamloops lawyer sent me a letter in Cloverdale where I worked for the Livestock Feed Board, to state that the repurchase agreement still was not properly registered and couldn't be until the Brysons signed off on it once again. By this time I was pretty ticked off and told Hook's lawyer to pay me $25,000 and we would sign over the repurchase agreement.

I didn't hear back from them for several months until Hook himself phoned me from his lawyer's office in downtown Kamloops. He was very nearly in tears on the phone as he explained to me that he had run out of luck on the ranch and was almost broke. This tale of woe affected me as Hook had taken the trouble to visit Dad now and again at his little ranch at Monte Creek and had become a kind of a friend. Knowing of this relationship with my dad, I reluctantly agreed to sign the lease, to Hook's obvious heartfelt relief.

Unbeknownst to me, it appeared as though that signing cost me and my siblings a lot of money. I found this out a couple months later when I was invited to my aunt's 90th birthday party, which was held at the home of her daughter. After visiting awhile their son arrived and said to me, "I hear Empire Valley has been sold again." I replied in the affirmative and inquired how he knew about it. "Oh" he said, "the timber company I work for were involved in that deal which saw them end up with the ten thousand acres of timberland.

"Uh oh," I said, "How much money did I leave on the table?"

"We were prepared to go to at least $50,000 to get you to sign," he said.

Ouch that hurt!

Chapter Four
Our First (and Last) Beef Drive

In the fall of 1956—after purchasing Empire Valley in August with a down payment of $75,000—we also had to make the first mortgage payment on the ranch of $75,000 to Henry Koster as part of the deal negotiated by my father. Henry, for his part of our agreement, had agreed to stay on at the ranch until the cowherd was moved back from the summer range to the bunchgrass winter ranges along the Fraser River.

Henry had anticipated selling the ranch and had speculated on the purchase of a thousand steers to fill out his range permit and perhaps his bank account. So when all the thousand steers and the rest of the cattle had been accounted for at the home ranch, we put together a herd of beef to sell to cattle buyers from Kamloops. Henry told us that he had a standing offer from his buyer (who knew the steers) for 17 cents per pound. However, Dad wanted to use the services of the BC Livestock Association and they sent out only one buyer. The deal we arrived at with that cattle buyer was 15 cents per pound for the steers, delivered and weighed with a three percent shrink to the rail yards in Ashcroft. This system was normal for a deal like this where the cattle would have been weighed at the rail yards with a belly full of grass.

Unfortunately, our original ranch purchase deal with Henry Koster had valued the steers at 17 cents a pound on the hoof at Empire Valley, which is the price he knew he could get from his Kamloops buyer. Obviously we were going to take a big financial loss but we had no choice because we had to come up with the $75,000 first mortgage payment one way or another.

The Empire Valley road to Canoe Creek across the Fraser River would not allow for us to truck the cattle out even if we had the funds to pay for the trucks—which we didn't—so off we went with probably one of the largest cattle drives ever seen in that country before and certainly since. Getting the herd across the Gang Ranch Bridge on the Fraser was our first and toughest test. The road switchbacks down to the bridge from the Gang Ranch/Empire Valley side, and then switchbacks up the other side onto the Dog Creek Indian Reserve. It then follows a very curvy, twisty road through Canoe Creek and Southeast towards Clinton. The Gang Ranch store used to sell a wooden plaque concerning that road which read:

> "Winding in and winding out,
> The only thing that is in doubt,
> Is whether the fool who built this road
> Was going to hell or coming out."

During this large cattle drive, I was in the lead with about fifty head of steers, and the rest of the cattle were being pushed hard on my tail by the rest of our crew. Try as mightily as I could, I couldn't get that first bunch of steers over that bridge to act as leaders for the rest of the herd. My job was not helped by a half dozen truckloads of hunters who had assembled across the bridge on one of the switchbacks where they were waiting impatiently for the cattle to clear the bridge.

After watching my efforts for a time they commenced shouting, "Drag one out onto the bridge." The effect of their loud commands made it virtually impossible to get these spooky animals across the bridge.

"Be quiet," I called back, "Or I'll come over and quiet you."

However, they kept hollering, "Drag one out on the bridge."

And so, in my consternation, I broke through my herd, leaving the crew to hold the entire herd against the bridge while I galloped across the bridge and up to the group of impatient hunters. You've never seen such a bunch of meek and now quiet hunters, who broke back up the road to their trucks with me on foot leading my horse and following them along slapping my quirt loudly against my chaps. I just wished my herd of steers was as pliable. Once I had the hunters all quieted down and patiently waiting in their vehicles, I returned to find the cowboys had managed to contain the herd at the entrance of the bridge. The steers were no match for me now as my temper tantrum was too much for them as well. Once the first few broke onto the bridge the rest followed in behind. The old bridge swayed and bounced three feet up and down in concert with the dance that 250 head of cattle at a time would do in unison as they poured across the bridge. As the suspension bridge heaved up and down those two or three feet per bounce, I think I was as scared as I have ever been.*

From the Gang Ranch Bridge we made it through bunch grass and sagebrush covered ranges to Canoe Creek, the headquarters of Jack Koster's BC Cattle Company. Jack was Henry Koster's older brother and a friend of my dad. Dad had arranged for us cowboys and cattle to be kept overnight at Jack Koster's headquarters in true cowboy fashion. The next day we made it to Indian Meadows, where the Gang Ranch had a large fenced holding pasture. The pasture fence had not been used for years and in many places was flat on the ground. The needed repairs to the fence made a long day even longer, but the cattle were tired and quickly bedded down for the night. Again Dad had arranged with the Tressiera Brothers Logging Co. for us to spend the night in their loggers' frame shack, which was ensconced within the boundaries of the holding pasture. The Tressieras were an old time Cariboo ranching and logging concern.

We cowboys were dead beat and quickly flaked out in the back of the two-room cabin, while the Tressieras and their logging crew played poker in the kitchen until midnight, with the aid of a gas lantern. Along about midnight the gas lantern ran out of fuel and one of the Tressieras attempted to fill the hot lantern with white gas which then unexpectedly exploded. Immediately the cabin was filled with flames and everyone ran for the one and only exit. We cowboys were right behind, all except Walter Grinder, whose modesty would not allow him to exit the cabin in his long johns because the cook was female. So it was rather anti-climactic to see Walter bouncing out of the flame filled doorway with his sleeping bag clutched around his middle, much like a runner in one of those old time sack races.

The next night we made it, less a few personal items burned in the fire, to Rocky Springs, one of Jack Koster's cow camps on the road to Meadow Lake. The following day we got as far as the Copper Johnny corral, a holding pasture belonging to the Empire Valley Ranch. Here, Baldy Boyd's cattle trucks picked up the herd for the final leg to Ashcroft. Baldy Boyd was from Clinton and he was a famous character in the Cariboo.

Boyd's trucks spelled the end to our one and only cattle drive to Ashcroft. The steers didn't net enough to pay that autumn's required $75,000 mortgage payment that we had negotiated with Henry Koster, so we returned home to pregnancy test the cowherd in order that we could sell any dry cows and finish off the payment. We did this with some difficulty because all of the cows had to be put through the chutes to be pregnancy tested by Lorne Greenaway, our veterinarian from Kamloops.

What difficulty we had, sprang from the fact that Henry Koster had left all the horns on his cows to better protect them from wolves, which were preying on his cattle in winter. Henry had managed to poison off the wolves with arsenic loaded bait with the help of the Game Department, but he had done nothing with the sabres on those cows and bulls that we had inherited. One of those sabre horned bulls had gored Henry Koster's wife Frances's favourite saddle horse in the Dry Farm Pasture the year before we bought the ranch, so we were quite timid putting those old sabre horned girls through the chutes. More than once we would go flying over the corral fence just ahead of a pair of switchblade horns.

When the cows were trapped in the squeeze, as well as pregnancy test, we cut off the tips of most of those scary horns. After the smoke had cleared and the dry cows had been shipped we were left with five hundred head of pregnant cows with which to rebuild our ranching fortunes. After shipping the dries—or empties—we trailed the remaining pregnant cows to the Big Churn Creek winter ranges, following the advice of Henry Koster before he had left the Empire Valley Ranch permanently. The Big Churn Creek flats were, and probably still are, the best piece of winter bunch grass grazing in BC, and all would have gone well had we not had one of the worst winters ever to hit the Cariboo. The winter of 1956 was a real monster from a frigid hell. By Christmas we had two feet of snow and temperatures in the minus 30 degree Fahrenheit range. On checking the cows just before Christmas, I discovered that they had abandoned the bunch grass flats because they were not able to graze through the two feet of snow covering the grass. They had escaped by working their way to the timber-covered hills south of the winter range. Because we had let most of the crew go for the winter in the interests of saving money, I was left only with Art Grinder to cowboy them home with me.

Art was a great young guy and one of the best cowboys you'd ever find. He was twenty years old in the winter of 1956. The Grinder family from Big Bar on the Fraser, all cowboyed for Empire at one time or another. They also contracted to build cabins and haul hay bales. The oldest was Johnny, then Henry, then Walter, and finally Hector and Art. Every morning at seven, Art and I would saddle up and leave the ranch in the dark to ride the fifteen miles over to the area to where the cows had scattered. This was not easy and each day conditions got worse with drifting snow wiping out our previous day's tracks.

Art and I rode for the month of January, rescuing as many cows as we could. We would take turns bashing a trail from the ranch through the drifted snows of the New Pasture, now called the Maytag Pasture—to the timbered slopes and ridges above Churn Creek. Once there, we would split up and track cows which were moving from clump to clump of dry pine grass exposed underneath the fir trees. In this manner we would each pick up ten to twenty head of cattle by mid-afternoon, and then we would head them for home and hayfields. We would arrive back at the Bishop Ranch hayfields about seven in the dark, cold evening, often at around the same time. The cold wind and drifted snow made for a very long, cold, frostbitten day.

As the days went by, the cows we found were becoming weaker and weaker on their starvation diet of snow and bleached out pine grass. There are almost no nutrients in frozen pine grass and very little water in dry snow. Following behind the little bunches of emaciated cattle slowly placing one foot ahead of the other made for agonizingly slow going, so we would walk and lead our horses while slapping our arms against our sides to stay warm. Listening to the starving cows' teeth grinding across their parched mouths is not a pretty sound, but somehow or other Art and I held on for that month and rescued most of our once proud herd of fat cows. I figured that during that month the temperatures never got warmer than minus 30 degrees Fahrenheit, and at times would hit minus 45.

You couldn't stay warm on the back of a horse, and even walking behind the cows in the wind and drifting snow was a cold hell. Unfortunately, the worst was not over for the starving cows. Once we had them shoved into the Bishop Ranch hayfields, we still had to find nutritious feed for them.

Whether because he was going to be leaving soon, or that the hay crop just wasn't up to scratch, or that Henry figured the cow herd could winter out on the Big Churn flats—whatever the reason—he had harvested very little hay when I arrived in August 1956. With a skeleton crew that included old one-eyed Louie Seymour from Canoe Creek and Harold Perkins from Big Bar, we had put up all the hay that was standing, and that was not very much.

We used teams of horses pulling hay mowers and dump rakes to get the hay cured in windrows. We did have one old, cantankerous, tractor-pulled hay baler, which I used to put up about 150 tons of hay when we should have had 500 to 600 tons—normally in this country you figure that one cow requires one ton or two thousand pounds of hay per winter.

We reserved any good hay for wintering our calves after weaning them from their mothers. Because we had such little good quality hay, we were forced to feed our starving cows from old stacks of rotting swamp hay that had been put up at the Bishop ranch in years past. This swamp hay was entirely insufficient in nutrients for our starving cows and probably made things worse, but we had to feed out this poor hay, for by spring we were out of any and all feed and had no resources to buy or haul good hay to them. In combination with the losses on the winter ranges and on the feedlot, we were now down to a nucleus of 350 head of cows from our bare bones nucleus of 500 head. That same winter our neighbour the Gang Ranch lost five times that number. Fortunately for our cows, spring came a little earlier than usual and we were able to get the cow herd out onto the Fraser River spring ranges for the first new green grass.

And so, wounded to the knees, we began our dispirited journey in 1956, one which ultimately ended in 1967 with Empire Valley being recognized as one of BC's preeminent ranches. We considered ourselves to be pretty good ranchers, despite our humble beginnings. Maybe this all happened "as much by good luck as by good management" as my mother used to say, but the fact remained that we had ultimately survived bad winters and terribly low cattle prices to build a magnificent ranching operation.

It must be said that between the years of 1956 and '59 we were stretched to the limit simply to pay our operating bills and the annual mortgage payment. I wonder how many ranchers today could get by with as little money as we did, for in those first years at Empire we had to "cut our coat to fit our cloth" as the old saying goes. The Bank of Montreal in Clinton would only allow us six thousand dollars per year for our operating loan and, our mortgage payment, after that first year's payment of $75,000, was $17,500 per year.

Getting by with cattle prices that were in the ten to fifteen cents a pound range for dry cows and twenty to twenty-five cents a pound for steer calves required us to be extremely parsimonious. We tried to muddle through and save all our heifer calves to rebuild our sadly diminished herd. I recall that we would make one seventy-five cent phone call per week to Robertson Brothers General Store in Clinton to order groceries on credit, payable monthly. These groceries were delivered by the Gang Ranch Stage Lines (owned by Sid Elliot of Clinton) to our Churn Creek mailbox, twelve miles from the ranch headquarters, once per week. Wages were four dollars a day for cowboys and farmhands, which certainly can't be compared to the two thousand dollars per month paid today, but it was all we could afford. Four dollars a day plus room and board was pretty common in those days. Cowboys were a lot tougher and more competent then, in my opinion. Those were tough, but character building, times and I miss them still.

Many ranchers did not expect us to survive that first winter, but by dint of hard work, good management and severe belt tightening we managed to confound the naysayers. As the saying goes - "When the going gets tough, the tough get going."

*Back Cover Painting by Peter Ewart likely portrays Mack after crossing the Gang Ranch Bridge with that big cattle drive.

Chapter Five

The Cowboy Takes a Wife (or Mack Got Married & Lost the Cows)

When I left for UBC in 1954, I was twenty-one and not entertaining any thoughts of marriage, but as fate would have it, things change. I think it was one of those fortuitous sets of circumstances that finally allowed me to register at the University of British Columbia in the Faculty of Agriculture. I had graduated from Britannia High School in Vancouver, along with my sister Donna, in June of 1949, when I had turned sixteen years old. From that point it seemed as though I was never going to get to university. First of all, I could not register at UBC at age 16 in those days. Secondly, my family and I had not the wherewithal to afford my going off to university until friends Betty and Joe Jamieson offered me room and board at their house near UBC if I would help with their three children. Joe was often away from home, running his construction crews.

We managed to scrape up the $300 tuition, and I was off to UBC. Unbeknownst to me on my way to meeting my future wife. Of course, she didn't know it then either, but what came next must have been meant to be. I was strolling down the halls of the UBC Agriculture building one day when I happened upon this beautiful creature.

In an instant I knew that this was the woman I wanted to marry. Her beauty and the way she carried herself as she walked by me caused me to instantly fall in love. There was a problem though, for it turned out that she had a boyfriend from Ontario, a fellow Aggie student named Radcliffe Weaver.

Elizabeth Stevenson had met Rad the year before when she was in first year Arts. In her second year, she decided that she would enrol in the Faculty of Agriculture in order to become a dairy bacteriologist. Liz and Rad were very close but I did manage to get an introduction to Liz when I dropped down out of the rafters in front of the pair while I was helping to decorate the UBC Armouries for the annual Farmers' Frolic.

Evidently she thought I was a bit of a show-off at that time, and she was probably right. Liz and I socialized because we belonged to the same group of friends who attended parties and functions together on weekends. As a matter of fact, those same friends are still some of our very best friends. Five couples of us were married within four months of each other in 1960, and we were all there to celebrate 50th anniversaries together in 2010.

Liz and I also socialized a little when she ran for—and was elected to—the Social Events Committee of the Agriculture Undergraduate Society and I ran for—and was elected to—Vice-President. As it turned out, I never served my term as VP because my family and I bought Empire Valley Cattle Company in 1956, and that ended my first exposure to UBC. Liz did not serve either, for she left Agriculture to enrol in the UBC College of Education with the intention of becoming an elementary school teacher.

Thus ended, it seemed, my first romance, but for my part, I was still in love. Our paths crossed again in one of my sojourns to Williams Lake and I began a writing campaign to interest her in pursuing our relationship. By this time she had been granted her teaching certificate. She taught in Vancouver for one year and then left for Hamilton, Ontario to teach there.

Hamilton is where Rad and his family lived and Liz had become very good friends with his sister Myrna, who was also a teacher. Myrna and Liz rented an apartment and lived together while they were teaching. By the end of that year her relationship with Rad had broken down.

My patient entreaties had borne fruit after all because when she eventually came home to Williams Lake to teach, I was able drive in from the ranch to arrange a date with her. At the end of each date I would ask Liz to marry me.

"I'll think about it," she'd say.

I grew so used to that reply that the last time I asked—the time she said *yes*—I just kept on talking until her answer suddenly caught me by surprise. Her one problem was that she was not sure how she would manage living on a ranch so far from town.

It was not long before I was standing before her stern-looking father, Doug Stevenson, his evening paper held in front of his face while he read. I interrupted him to ask for his daughter's hand in marriage. Time stood still for a moment while Doug pondered that, and then he smilingly gave his acquiescence. Before long a date was set; Liz and I were going to be married in July.

**Mack and Liz wedding on July 4th 1960;
(credit to Fred Waterhouse)**

An interesting side note—prior to me popping the question to Liz, it came to pass in 1958 that I was attending the annual Bull Sale in Williams Lake and had gone to the Bull Sale Dance at the Elks' Hall. It was there that I asked a young lady to dance only to find out, later that evening, that she was in fact Liz's younger sister, Rhona. I confided in her that I would someday marry her sister. Rhona evidently got such a kick out of this confidence that she told her family that she had met some cowboy at the dance who'd said he wanted to marry Elizabeth. This was greeted with chuckles by all, for after growing up in small towns all her life, Liz loved living in the big city. Of course, as it turned out, I would have the last laugh on that score.

Anyway three weeks before the wedding, in June of 1960, seven of us set off—complete with a pack string, spare horses, and over 350 head of cattle—and headed for the Gun Creek range that Don Gillis, Walter Grinder, and I had discovered in the summer of '58. Our crew naturally consisted of Don, Walter, Francis and I—plus my sister Donna, brother Dunc, and a couple of capable cow hands. All went well from our headquarters until we started moving the cattle into unfamiliar country. That's when the work began in earnest, and we had an ongoing fight with cows that definitely did not appreciate the potential for lush valleys that lay hidden away beyond the mountains.

Cattle drive to Fish Lake; then onto Gun Creek range

I was riding my dad's mare, Dude, on the fourth morning out, when two heifers broke back and headed for home. I gave chase and eventually cut them off with Dude, but one swerved back under the mare as we attempted to turn them. We were going flat out when it happened, half a mile behind the main herd. Dude went end over end, and I piled head first into a shale rock sidehill. Some time passed before Don Gillis came back to look for me and found me out cold in the rocks, with Dude waiting patiently nearby. My recollection of the next three or four days are rather hazy as Don and the others took turns keeping me awake because I was in shock. Don's dad was Dr. J.J. Gillis of Merritt, and so he thought he knew all about shock.

I wasn't much help to the crew when we arrived and camped at Fish Lake with a herd that was becoming increasingly disenchanted with the idea of going further west. That night we built a corral in a U shape surrounding the cattle. This left the cattle with Fish Lake on one side of them and a brush corral surrounding them.

It soon became clear that is wasn't going to hold the cattle, so about midnight we built fires all around the makeshift corral to keep the cattle inside. This worked until some particularly crafty cows led the herd into the lake where they swam across a narrow strait and into the rocks, windfalls, and second growth trees abounding on the far side of the lake. The rest of the night was spent with our crew on foot, chasing the herd back down out of the rocks and windfalls and across the lake and into the corral. I suspect that I was still too far out of it to have been much help in this endeavour.

We persisted and over the course of the following days managed to drive these reluctant pioneering cows out to the Tyax Valley, about 75 miles from home. By this time the Tyax Valley was beautiful and lush, the countryside was filled with deer, and the air was fragrant with purple lupines. The Tyax River, however, was in flood, so we couldn't cross the cattle into the Spruce Lake country.

We left the cattle on the Tyax, under the care of Francis Haller and headed for home. Francis built a tent camp at a spot he named Bannock Camp. Francis was an expert at making bannock, a flour based Indian concoction and a cowboy staple. Later he packed metal wall siding from the old deserted Manitou Mine, near where the Tyax River joins the Relay River and the Mud Creek. With this metal siding he constructed a pretty fair cow camp style metal hut, roofed with jack pine saplings covered with tar paper and dirt.

Mud Creek cattle drive

A couple of years later, that same metal hut was to be the scene of a scary evening indeed. We were all camped at Bannock, having brought out another herd of cattle for Francis to tend. Francis and Walter Grinder were sleeping in the metal hut while the rest of us cowboys, including my sister Donna, Don, and their kids, Monica and John, were camped in tents nearby. About midnight we got hit with a heck of a lightning storm. I was lying awake, counting the seconds between flashes and the following thunder claps; thunder one second after the lightning shows the lightning to be hitting a distance one thousand feet away—10 seconds equals ten thousand feet away. When my count indicated that we were dead centre of this fantastic lightning storm, hitting trees all around us, I hollered at Francis and Walter: "Get the hell out of that metal hut, you two, before you are electrocuted!"

They responded that they were too afraid to touch the metal door, for sparks were flying all around. To this day I don't know why they weren't fried in that metal hut on the exposed side- hill.

But back to my story and 1960. Having recovered fully from my head injury, I had a week to prepare for my coming July 4th nuptials to Elizabeth Stevenson.

Her father, Doug Stevenson, operated Mackenzies Ltd., the department store in Williams Lake started by Liz's maternal grandfather, Roderick Mackenzie, in that cow town. As an aside, during the depression, Roderick Mackenzie was able to help many of the Chilcotin ranchers survive with their ranches intact as he did not ask for payment on their accounts until the ranchers had had the opportunity to sell their cattle at the fall cattle sales.

Anne, Liz's mother, was the daughter of this pioneer merchant—and one time MLA for the Cariboo. Doug was a retired mining engineer, and Anne was a high school English teacher. Both Anne and Doug were popular pioneers in their own right, and so when July 4th rolled around, the town basically pulled down the blinds, closed up the doors, and came to the wedding.

As a result, our wedding was quite a big social event; we had five bridesmaids, five groomsmen, and half the Cariboo in attendance. The bridesmaids were all in white, carrying white satin muffs festooned with different coloured flowers. My groomsmen were mostly big, husky characters, like my brother Duncan—my best man—and brother-in-law Don Gillis. Dressed in white dinner jackets they were having a bit of a lark meeting all the townsfolk and crunching the hands of men they met in the reception line.

When Big Dan Lee from Hanceville (Lee's Corners) joined in the receiving line at the Lakeside home of the Stevensons where the reception was held, I whispered to the boys to pass the word along to really crunch Dan's hand, for though he was huge, he was a wimp. Much to their consternation Dan rendered them pretty much left handed for the rest of the day; Dan immediately recognized what they were up to and crunched each of their right hands in turn.

In truth I had previous knowledge that Dan had this pair of giant sized vice grips for hands, for a few years earlier, he and my dad had been in a partnership with a herd of cattle on our family owned ranch at Pavilion. Dan had come down from his ranch in the Chilcotin to help us brand, castrate, and dehorn calves. My dad had given Dan a new pair of Burdizzo castrators that would be easier on bull calves than knife surgery. His instructions were to close the pliers on the scrotum and put lots of pressure on the handles of the "metal" pliers to crush the Vas Deferens without surgery. On the first bull calf, Dan did as he was told; squeezed hard and broke both handles off the Burdizzos. So Dad had to go back to the knife once more for castrating.

Anyhow, Liz and I were married but we didn't get to stay for the following party because in those days the newlyweds headed right off on their honeymoon. We missed all the hijinks but there must have been plenty, because a few weeks later, coming back from the Oregon Coast, we stopped in at E.A. Lee's in Vancouver to pay for the rented white dinner jackets. The attendant presented us with an unexpectedly monstrous bill, and when I queried him as to why the bill was so high, he said that the white jackets all had to be remade and one could not be repaired. My schoolteacher cousin, Doug Brett, was one of the groomsmen; after spending a year teaching school in Telegraph Creek, he was returning to his home in Vancouver. It was he that I had asked to take the tuxes back. Apparently he had made up a little story to explain their disastrous condition by telling E.A. Lee that he had brought them down in the back of a cattle truck. We managed to get the bill reduced but we did have to purchase one of the white jackets. We still have that jacket lurking around somewhere. Our kids used to use it for dress up at Halloween.

Sometime after we returned from our honeymoon, Liz realized that she might be pregnant. As she was slim, Liz showed signs of her pregnancy quite soon. As was the custom in those days, people would quiz Doug Stevenson when they came into the store, as to when the baby was due. I think they were hoping there would be a little scandal here, after that big white wedding. Nowadays people could not care less but that was not the case in 1960. Doug's instant reply to the query was to say, "The baby is due in April, but don't bother counting, it will be nine months and fifteen minutes after they left the wedding reception!"

This would have meant we had barely made it out of Williams Lake, when in fact, we had gone all the way to Kamloops. Liz was concerned that the baby might be early, which would have given all the wags something to snicker about. Fortunately, baby Lisa Jean arrived on April 14, 1961—nine months and ten days later.

We discovered another incident from the wedding when we returned from our honeymoon. Most of my friends and relatives had stayed in a motel in Williams Lake. Some of the more exuberant youngsters had managed to run through a screen door on the motel unit. I knew about this and offered to pay the owner for the damage. He said $100 would cover it, and I paid him with a ranch cheque and then went off on our honeymoon. It wasn't until later, in talking to Don, that I discovered that he had also paid the motel owner $100. Some months later, when I mentioned this duplicate payment to my father-in-law, he laughed and said, "What do you know, I paid $100 for that screen door as well."

Liz and I returned to the ranch after two idyllic honeymoon weeks on the Oregon Coast. I plunged right into helping Dunc and the haying crew put up baled hay for the winter. This kept me at home, for I didn't want to leave my bride and head out into the mountains if I could put that off for a while. The haying crew got quite a chuckle out of me hanging around the home ranch, saying Liz was "sleeping on my shirttail," an expression she definitely was not fond of.

It came as a bit of a surprise when Francis Haller attempted to ride in from the Tyax to tell us he was having trouble keeping track of the cattle. I say attempted because he had a heart attack at Yodel Camp, twenty-eight miles from the home ranch. When Francis attempted to saddle up Whitey[2] for the last day's ride home, he suddenly ended up on the ground. After passing out for a while, he crawled back to the cabin, and with some difficulty, melted a pound of lard on the stove, thinking he had poisoned himself, and that the lard would contain the poison. He drank this potion and after a long rest he managed to somehow saddle up the grey mare and pull himself aboard. He knew he would probably pass out again, so he tied himself to the saddle with the saddle strings. He then let the old mare have her head on the way home to Empire.

Whitey was a fast, smooth walker, so Francis swayed and swooned all the way to Hog Lake, a few miles from home, where my mother happened to be showing some "rock hounds" the thunder eggs (agates) which abounded in that area. At Hog Lake they encountered the grey mare carefully packing Francis home. Francis was semi-conscious, but survived the quick car trip to Kamloops and a stay in the hospital. That was the first of two heart attacks for Francis.

The second one occurred at Mud Lake. That one occurred at night in his tent, and Don and I took turns keeping the old boy alive all night. Long before daybreak, Don saddled his horse and headed for the Manitou Mine and from there down the road to Tyax Lake so that he could phone-call in a Medivac helicopter. Francis' heart was fibrillating when the helicopter arrived. Again he survived. Later, when he got back to the ranch, I suggested that he not go back into the mountains. He was not having any of that and said that the only way he was going to die was, "In the mountains with my boots on."

Anyway, right after that first heart attack and the pound of lard antidote, Francis indicated that a lot of cows were missing on the Tyax, and since Francis couldn't go back for some time, Don saddled up his string of horses and headed back to find the lost cows. I begged off, being the "Honeymoon Boy," and all that.

That state of repose didn't last long though because Don was back at the ranch in a week, saying that he couldn't find a "bloody cow." That really shocked me, and I realized that the honeymoon was definitely over. Again we shod up a string of horses and headed for the Tyax. Our first day there we found some fresh tracks leading up into the high country and an alpine range that we dubbed the New Range.

This was where we discovered most of our missing herd. They had become quite wild, not having seen any humans for some time and were being harassed by bears that were after any calves from the late calvers. On the first trip up, following tracks into the New Range, I could hear the faint, distant bawling of a calf in distress.

Don and I hustled along until the bawling was closer, but becoming more muffled. We broke into a clearing where a small bunch of twenty or so cows was peering into the base of a big spruce tree with branches that extended all the way to the ground. The calf's laments were becoming quite muffled, and so I realized that a bear must have him under the spruce tree. I piled off my faithful horse, Shifty, and pulled my .30-30 out of its scabbard. I parted the branches of the tree to see inside and was immediately face to face with a huge black bear straddling a month old calf.

Immediately I pulled back, stuck my .30-30 in its face, and pulled the trigger. It was all that Don's horse and my Shifty could do to drag the big black bear out from under the tree and off the calf. Miraculously the calf was still alive and ran to join its mother and the whole bunch vamoosed down the trail. Don and I were amazed to find out that our very sharp pocket knives were barely able to cut through the hide of that big fat blackie. Our guess at the time was that it weighed in the neighbourhood of five hundred pounds. He was the biggest black bear I had ever seen and had obviously grown quite fat on his predations.

We managed to cut out a five-pound roast from his rump for dinner. The roast came with a five-inch coating of fat, which we later rendered down and applied to our leather chaps, having heard that bear grease was great water repellent—how's *that* for Davy Crockett stuff? Much to our chagrin, we later discovered that whenever it rained, the bear grease seemed to act as a conduit which ran the rainwater right through the leather chaps to soak our pant legs underneath. That big bear had his last revenge.

This trip had an element of comic-tragedy too, for after the incident with the bear, Don and I took off after the bunch of cattle we had found surrounding the helpless calf. With some difficulty we managed to find and run the small herd downhill into a clearing on the main cow trail. On the way down, we had lost a few especially wild heifers, so I left Don to ride around the herd until they settled down while I headed back up to the New Range to see if I could connect with the runaways.

I did just that, and when I reappeared in the clearing a few hours later, I ran a half dozen head of steers and heifers into what I thought would be a calmed down and placid herd of domestic cows. Imagine my surprise when my little herd broke into the clearing and quite promptly disappeared out the other side of the clearing without breaking pace.

A shame-faced Don appeared saying, "Mack, you'll never believe this." Of course, having heard Don use this expression many times before, I had no trouble believing it.

It turned out that, after a while, most of the cows had laid down to rest, which gave Don the opportunity to do the same. Don had occasion to lie down on an old rotting log in the sun and soon fell fast asleep, only to be wakened quite rudely some time later with the feeling that something was crawling across his stomach. The crawler turned out to be a large Bull Snake, which had no reason to suspect that Don was deathly afraid of snakes. He jumped to his feet with a whoop and waved his arms dramatically. The upshot of this was one snake headed into space and one herd of cows headed for parts unknown in high gear.

So much for that day's work.

Over the next few weeks Don and I discovered most of the cows' new hiding places, and a few months later, after having the herd fattened up on the succulent grasses of the alpine ranges it was time to head them for home.

The majority of the cows hit the trail for home without too much trouble but it required me to go back several more trips to find stragglers. Usually I would fly the area with Langley Air Services when snow on the ground made spotting cows fairly simple, on the steep and windblown slopes of the Tyax River and its tributaries. I knew we were short cattle when I got a herd count late that fall after we had assembled the cattle on the home ranch to wean the calves and ship beef to market.

It was no surprise when, late in November, we had a telephone call relayed to us that hunters had heard cattle bawling near Big Dog Mountain just west of the Yalakom Range, approximately fifteen miles from where they should have been. I again assembled a crew and sharp shod horses for yet another rescue mission. With me went my sidekick Walter Grinder and his brothers, Henry and Hector.

We loaded two pack horses with 200 pounds of oats apiece and headed out. At that time, just before Christmas, there were about six inches of snow at the home ranch, just the right amount for winter grazing on the bunch-grassed flats of Churn Creek. However, when we got to Yodel Camp there was a foot of snow. Going over the higher Red Mountain Meadows, this became two feet of snow. Further west, at Fish Lake, there were three and a half feet of snow, and as we climbed up over the height of land that led down to Mud Lake, the snow was now even with the top of our horses' backs. First I took my saddle off and rode Dolly bareback to break trail. She would jump up in the air and land in the snow in front to crush out a trail for the rest of the string. After playing out all our spare horses in similar fashion, I took to bashing the snow down myself, in a similar fashion to Dolly. This soon played me out as well, what with my bulky winter clothes, chaps and all.

In this manner we made it to Mud Lake Camp and surveyed the possibility of tackling the now five to six feet of snow that we would have to conquer to climb through the timber out of Mud Lake in order to head up a steep trail to Quartz Mountain on our way to Big Dog Mountain, a place none of us had ever been before and well off our cattle range. With a great deal of remorse, we turned back and consigned thirty head of steers to their sure snowbound starvation.

Two summers later we were again headed out with a new bunch of cows and calves to tempt the use of this tough piece of range. We were five cowboys and five helping friends pushing five hundred head of cows. The first lap was relatively easy as my father, Dunc, and I had built sixty-five miles of four-wheel drive road, as far as the height of land entering into the Tyax Valley with the TD-9. The road itself was not built without a lot of hard work, but the skill of my father at building roads with a little TD-9 Cat would have to be seen to be believed. When my brother Dunc and I alternated, taking turns away from ranch business to chainsaw trees ahead of the Cat, I occasionally took a turn at the Cat. It always seemed that I took over at a time when we were surprised by rock and I just couldn't handle it. Dunc turned out to be almost as good as Dad and that was the end of my Cat work. Cowboys don't seem to be all that good at machinery, better at horses.

On this, our second cattle drive, things went well until we hit the end of Red Mountain Meadows and attempted to push the cattle off the meadows and down into Churn Creek. The cattle balked and so we camped for the night. Walter threw up tents, then hobbled and staked the horses. (To hobble a horse means tying the front feet together so that there is just enough slack that the horse can graze but not run away home.) The next morning we were up at 4:30 and headed back down the meadows to pick up the herd which had retreated three or four miles between sundown and sunup. Again we put the pressure on the cows to get them through the bogs and jackpines at the end of the meadows. Francis Haller, who was designated cook, left his camp to join in behind on foot, thumping a dishpan. In this manner we eventually emerged down onto Churn Creek and finally got the cows as far as Fish Lake where, this time, they didn't fight as they had the year before. They were remembering the lush green grass that awaited them at the end of the trail.

Dad was following behind with the Cat, cleaning up the trail and rerouting the road around some of the swampy parts. We got to Fish Lake about two in the afternoon and decided we could make Mud Lake by nightfall and put the cattle in the big sheep herders' corral there. We were exhausted by the time we got to the corrals and had to wait on Dad who had gone back with the TD-9 from Fish Lake to pick up the camp supplies and the crews' personal items including sleeping bags, load them on the wannigan and haul it all up to the Mud Lake corrals.

He clattered in at 11:00 PM with some food, some blankets, and a disastrous tale. It seems Francis had left fires untended while he thumped the wash basin along behind the herd. Somehow these fires had jumped sparks onto the tents and bedding and from there to the spruce forest surrounding this last meadow we had camped on. Dad and Francis arrived back to the camp on the Cat to find an inferno of exploding spruce trees and a devastated camp. Fortunately a few of the sleeping bags and most of the food had been piled by the cowboys away from the tents before we left early in the morning

One rolled up tent had been saved, so at Mud Lake, for sleeping purposes, I requested that those without blankets should lie down on this tent and I folded the other half over top of them. This was done at midnight. For myself, I found an old, dried up moose hide, and made a crude bed under a big spruce tree. Most of us, except the few who had blankets, including my sister Donna and her husband Donald, spent an awfully cold night, with the temperature at freezing. At 4:30 the next morning, Marvin Tenault, from Indian Meadows, wakened those who were asleep with a cold lament, "Let's get the show on the road. I'm freezing."

After two more days of hard herding and poor sleeping, we again left Francis in control of the cattle at Bannock Camp. We had intended to get about 250 head across the Tyax at its confluence with Spruce Creek and from there into the Spruce Lake/Gun Creek Range. However, the river was too high, so we pulled back with the intention of bringing another drive out from home and then pushing all the cattle across the Tyax when the river retreated early in the morning.

Don Gillis, carrying his dog, after crossing cattle at low water early in the morning

Later that summer Francis appeared at the ranch for a little R&R and announced that he had seen three or four of the lost steers that must have survived those harsh temperatures and the horrific snowfall two winters before. They were as wild as deer, but Francis intended to keep working gentle cows around them until they became accustomed to humans and were docile once more. In this manner he succeeded in getting two large three-year-old steers home that fall. They were huge in relation to their yearling cousins.

We managed to work these into a beef shipment trucked to the railhead at Ashcroft and thus recouped a little of our loss from the thirty head that we had to abandon two winters before. The following year Francis got another one of the lost steers into the herd and home. This steer had survived his third winter on the wind-swept side hills of the Tyax and—a bigger feat—had eluded the hungry grizzly bears. However, we never did get the fourth steer, so it was a really big surprise when, the following year, I got a phone call from Tuffy Derrick, the manager at the BC Livestock Auction Yards in Williams Lake. He told me that the Brand Inspector had discovered one of our steers at the auction sale. The big five year-old steer no doubt stood out at the sale and caused the Brand Inspector to question the ownership of the animal for he couldn't make out the brand on the steer's thick winter coat until he clipped the steer's hip hair. One Chilcotin rancher was obviously dismayed at not getting the sale of this valuable steer.

Somehow the big steer had gotten out of the Tyax, up through the Tyax pass, and down into Graveyard Valley. From there he probably followed Big Creek to one of the smaller ranches west of Hanceville in the Chilcotin. The steer, according to Tuffy, was the biggest steer ever to be weighed through the scales at Williams Lake, topping the scales at around two thousand pounds. The cheque that we subsequently received totalled $370.00 at a time when prime, eight hundred pound steers were fetching about twenty cents per pound on the hoof—today it's more like a dollar twenty per pound. In retrospect, we lost a lot of cattle partly on account of my marriage to Elizabeth and my subsequent reluctance to tend to cattle business. In the end nobody was hurt and Liz and I survived this and many other of life's little skirmishes that we always managed to find a way to weather.

**Mack and Liz's 50th Wedding Anniversary
July 4, 2010-- photo taken by Laurie Marchand**

When we left the ranch, I took Whitey to Kamloops with me. My buddy Nick Kalyk and I, along with a friend, Kamloops lawyer and fellow Liberal Jarl Whist, bought a little 150 acre property outside of Kamloops, in Barnhartvale. Nick and his wife Margaret built a house on the ranch and Whitey became a pet of their girls, Julie, Leslie and Valie. Margaret would often look out the window to see the girls hanging off Whitey's mane. One day, to her shock, she saw Whitey happily eating grass with the girls, sitting underneath her, having a picnic. The most surprising incident in Whitey's time on the Barnartvale Ranch was the day our neighbour, Rick Walley was passing the field on his way to town. He came bursting into Nick and Margaret's house.

"Whitey has a colt!" Unbelievably, there was "Old" Whitey, standing in the field, tenderly caring for her new little baby. She had a white star on her forehead so, of course, the girls called her "Star." We had thought Whitey was much too old for having a colt and we don't know when she was bred or what stud impregnated her.

Chapter Six

Shepherds & Thanksgiving

Does one Thanksgiving holiday stand out for you as particularly unforgettable? I know that one does for me, without a doubt.

First a little background—In the summer of 1958, Canadians everywhere were closely following the career of John Diefenbaker, who was bursting onto the national political scene as Canada's Conservative Prime Minister. He was determined to advance Canada's role on the world stage. When Don Gillis, Walter Grinder and I went scouting for additional summer range for our rapidly expanding cow herd in the summer of 1958, we were not interested in political events. Our ambitions were to find more grass for our hungry and rapidly growing cow herd. That meant extending our own grasp over a huge wilderness area that was being vacated by the famous Hayward Sheep Ranch of Kamloops—Canada's largest sheep ranch at the time. You see, the back end of the summer range at Empire Valley runs clear to the Gun Creek/Spruce Lake country north of the quaint little town of Gold Bridge which is near to the old mining town of Bralorne, once home to Canada's largest producing goldmine.

The Haywards were leaving because they were running out of shepherds willing to spend the lonely summer in the alpine grazing areas of Mud Lake, Paradise Valley, Prentiss Lake,

Quartz Mountain, Red Mountain, and Poison Mountain to name but a few of their sheep camps. The Hayward sheep herders were famous for the long trek they completed each spring from the sheep's winter home at Noble Creek on the Westsyde Road, north of Kamloops. These sheep would eat their way west to Cache Creek, and then north to Clinton along the Cariboo Highway. As a boy I can remember pushing nonchalant sheep out of the way so that my dad could navigate our truck slowly through the thousands of sheep chewing their way up the Cariboo Highway. The sheep's constant companion was a single herder and a half dozen Border Collies he directed with hand signals. Sheep dung greased every foot of the highway while the phlegmatic ewes stared down truckers and turned the Cariboo Highway into a traffic gridlock every spring, just like clockwork.

Perhaps that was part of the reason for the Haywards' decision to discontinue their annual migration to what would become a big part of the Empire Valley Cattle Company's summer ranges for cattle. After leaving Clinton, the Hayward sheep would have headed Northwest to Jesmond and from there down Big Bar Creek and across the Big Bar Ferry to the West side of the Fraser River. Imagine trying to load and successfully transfer thousands of bleating sheep across a raging Fraser River on a fifty foot scow ferry that hung from cables stretched across the river. That took guts and perseverance on the part of the ferryman and the herder. Once safely ensconced on the west side of the Fraser River at Big Bar, these woollies were herded into this huge wilderness area. Having crossed the river and left "civilization" behind, the herder now really earned his keep trying to keep the grizzly and black bear populations from fattening up on mutton stew and lamb chops.

I can recall being camped at Mud Lake and hearing the herder two thousand feet above us firing his .30-30 rifle all night at regular intervals to keep the bears at abeyance from his nervous flock. Whatever their reasons, the Haywards did give up on this incredibly beautiful and luxuriously verdant alpine grass country. This move allowed us to move in and make application to the BC Forestry Services Grazing Division for Empire Valley cows to range over an additional half a million acres. But before we could make any application to increase the size of our summer grazing range, we had to get out there and scout out any opportunity to make that application to the Forest Service's Grazing division. So that's what Don Gillis, Walter Grinder and I were doing out there in the country West of Gold Bridge.

The quaint little town of Gold Bridge plays a part in my tale about Thanksgiving. Gold Bridge happened to be the home of three sons of gold miners, who had gotten hooked on the beauty of the adjacent Spruce Lake country, perhaps because of Charlie Cunningham's stories and pictures. They all ended up working for Empire Valley at various times in those first years that we moved cattle into the country that the sheep had left. Their names were Barry Menhinnick, Gordon Gerrard, and Don Tremblay.

This Thanksgiving story concerns Don Tremblay and me. Don was a nice young man whose western ambitions had led him to take a farrier course—a horse-shoeing course—down in Montana, where he had also managed to pick up a Montana drawl. He had taken the course so he could prepare himself for the life of a packer and guide in the gorgeous wilderness country surrounding Spruce Lake. So it was natural for Don, Gordon, and Barry to hook up with Francis Haller as he babysat the Tyax, Spruce Lake, and Gun Creek herds during those early years of Empire Valley taking over those ranges. Today the Spruce Lake/Gun Creek Range is reserved only for horse packers but then it was a marvellous place for us to summer four hundred heifers and the bulls required for breeding purposes—one bull for twenty-five heifers meant around 16 horny bulls.

On one of his infrequent trips back to Empire Valley headquarters, Francis told me about these young men, and how he thought he could use and train them to be cowboys. Francis took a particular liking to Barry Menhinnick, but somehow or other we got all three to come aboard at various times.

Because he knew the country from packing into our recently acquired ranges, it made some sense to have Don Tremblay with me that particular fall for the gather at the back end of the summer range. We would normally start kicking the cows out of that Spruce Lake/Gun Creek/Tyax ranges in mid-September. This usually required three or four cowboys, each with his dog, and a string of three or four horses. We also needed a wrangler and half a dozen pack horses for carrying our supplies. Two such gathers would leave the country pretty clean, except for stragglers. After the main herd and cowboys had retired towards the ranch, I was accustomed to going in for one last look before flying the country in a Super Cub rented from Skyway Air Services in Langley, to spot the stragglers. On this particular last look- final trip, Don Tremblay had been available to come along. He was good company and liked to tell stories with a western drawl, so we got along fine as we searched out the hidden valleys surrounding these far-flung ranges.

After two weeks of steady riding, Don going in one direction and me in another, picking up a few cows here and there that we headed down the trail to home, we were headed generally back in the direction of the home ranch. Snow was in the air, our grub was running out and we were arriving famished back to camp each night. Our supplies were depleted to the point where we would soon have to head for home or face starvation. I convinced Don that we could cut up the mountain from our Mud Lake camp and use up the last of our rations to ride up to the Quartz, Buck, Poison, and Red Mountain ranges for a last look through there before we headed for home and grub. If we were lucky we would not find many cows and we could look forward to finding some cached grub that we had left at Yodel Camp, at the foot of Red Mountain, when we departed from the back country. From Yodel we could call it a day and ride the remaining thirty miles to home. We would certainly have an empty belly, but at least we would have given the country and any cows that might be remaining there our best shot before winter set in.

Mud Lake Corrals—cowboys heading home-cowboy camp to right, out of sight

Don reluctantly, and with much stomach grumbling, agreed. Riding up out of Mud Lake, we got into the Poison Mountain Cow Camp just after what should have been lunch on Thanksgiving Day, 1962. "You take Quartz and Buck Mountains and I'll take Poison and Red Mountains," I said, "and we'll herd anything we find down the trail towards Churn Creek."

Don agreed even though he looked pretty hungry and downcast. Now it chanced that my area around Poison and Red Mountains, was accessible by four-wheel drive truck, to hunters coming either from Clinton—across the Big Bar Ferry and up China Ridge to Poison Mountain—or from Lillooet to Mohaw and then up the Yalakom Road to Poison Mountain. As I followed the four-wheel drive road looking for cow tracks in the general direction of Red Mountain, I was attracted to what looked like a message of some sort, on a piece of cardboard, hanging from a tree branch. This was close to the recent remains of a hunters' camp set up near the road. Imagine my surprise when I rode up to the tree with the cardboard sign swinging in the wind, to discover three blue grouse and a half gallon of red wine had been attached to the same branch as the sign. On closer inspection I read this message printed on the cardboard beer carton—DON'T TAKE THIS UNLESS YOU REALLY NEED IT. LEAVE IT FOR SOMEONE WHO REALLY NEEDS IT.

I figured Don and I qualified, so I hefted the three grouse and wine into my saddlebags and continued on my way looking for cow tracks.

Having found no cows, I arrived back at the cabin about 6:30 in the cold and dark. I expected a downcast and hungry Don Tremblay would be waiting for me with at least a warm fire going in the stove but the cabin was cold, dark and empty. Don, despite his misgivings, was doing his job and was obviously giving the Buck/Quartz Mountain country a good search. I got a fire going, lit some candles, cleaned the grouse and made the last of the pancake mix into biscuits that I put into the oven. Sometime later I could hear Don pull in on Spooks and go about settling her for the night. Imagine the look on his face when he opened that cabin door to the steamy aroma within. "My God," he gasped, "What's that smell?"

His eyes were as big as saucers as he stared at me basting the three blue grouse in a pan on the stove. The table was all set, complete with candles sputtering out a welcome and fresh biscuits were in the warming oven. He pushed open the cabin door, hardly believing his eyes or his nose. In my memory I can still see him, standing there in his chaps and his big black hat, staring in disbelief as though I was playing some kind of mean trick on his senses.

"Pull up a chair, Don, pour yourself a glass of Calona Red, and I'll fill up your plate with this tasty turkey and biscuits. Sorry about the stuffing, I was a bit short on onions. However, the gravy and biscuits will make up for it."

Don needed no second invitation.

Never in the history of cowboying have two jaded, starving cowpokes enjoyed a Thanksgiving meal more than that one. This story makes a guy really understand the true meaning of Thanksgiving. I hope the hunters who left that cache on Poison Mountain get to read this and congratulate themselves on having given a couple of starving cowboys enough incentive to finish the gather before the real earnest winter set in. Don and I certainly gave thanks, toasted our benefactors and our good fortune in that lonely cabin atop Poison Mountain.

Chapter Seven

Things Not to Do with Your Bride to Be & Things Not to do with Your Wife

The fall of 1959 was my first hunting season together with Liz, and I elected to take my future bride out to see what the really good big game guides do for their hunters. Since I didn't have any hunters on this Sunday afternoon it seemed like a good day to take my young and beautiful fiancée to the big Churn Creek country where our guides were operating that day. I took along my short barrelled .30-06 Husquavarna that had been given to me by friend Jerry Ingleby—just in case we were attacked by a really mean buck. Along the way we spotted a coyote ambling through the bunchgrass and minding his own business. Here was an excellent chance to show my future bride why I felt I was the finest shot in the whole of Empire Valley and possibly the whole of British Columbia.

At a range of a hundred yards the coyote had little chance to escape from such a remarkable hunter. However, it was not to be—my showing off, that is—for after two of my best shots the coyote decamped unscathed over a rise, and I was left explaining to my fiancée how the light was bad and a running coyote is a very small target. Compounding matters further I had used two of the six shells I had brought along. The reason for so few shells was that a really, really good big game guide needs very few shells to establish his presence as an excellent shot.

We continued on through the Dry Lakes Pasture and up the hill and through the gate into the Big Churn Creek Pasture without any more ego busting events till we came out on the Big Churn Creek Flats and spotted brother-in-law Don Gillis running flat out across our path.

"Mack, am I ever glad to see you," he huffed. "One of my hunters wounded a buck and I have run out of shells trying to get him as he is running at a pretty good lick. I think I broke a back leg so he's definitely slowed down. I need you to give me hand to finish him off. You've got your rifle I hope."

"Of course, Don," I said, "No problem. Where is he?"

Donald gestured in the direction he had last seen the deer. I had Liz stay in the truck where she could watch all the action as these two great white hunters finished off this wounded buck. The buck had already lain down but when he saw us he jumped up and was away again. Dropping to a crouch I lined him up in my four powered scope and Bingo; much to my surprise the buck kept going. That was my third shot out of six, and I quickly burned off the other three shells with no better results than the first three.

I found out later that I had loaned my rifle to one of our guides who had dropped it but had not told me about it. The scope sight was drastically out of kilter. Try to convince my future wife of that.

"Uh oh Don, now I'm out of shells too. What shall we do?" I said.

Fortunately, I had my lariat rope in the back of the truck. I grabbed it and then Don and I went again after the wounded buck. The exhausted, wounded animal by now was in shock and only hobbling along. I managed to rope him over the horns and stretched him out. I then used the rope to hold him while Don found a big rock and clubbed the buck senseless so we could cut its throat. That done, we gutted him out and threw him in the back of my truck.

The look on my Liz's face as she took in all of these shenanigans was one I'll never forget. I'm sure it is also one she'll never forget for she never again was willing to go out with a really good big game guide to watch him do his thing. Certainly not a good way to impress your prospective bride.

THINGS NOT TO DO WITH YOUR WIFE

In late June of 1964, Liz and I were coming up to our 4th wedding anniversary. We usually took our farm crew and cowboy crew to Williams Lake for the annual Williams Lake Stampede, which was held the last week of June and took in the Canada Day holiday on the 1st of July.

We had been married on July 4th, when all the ranchers from the country surrounding Williams Lake had come to town for the Stampede and could stay on for an extra day or so for our wedding. The Stampede had been started by Slim Dorin and a few others many years before—the same Slim Dorin, who, as cow boss of Nicola Stock Farms, had trailed 500 head of cattle for us to feed at the Voght Ranch in Merritt

Complicating matters in June of '64 was the fact that a late snowmelt in the mountains had roared down Churn Creek and washed out our bridge, which was the only way vehicles could leave the ranch. This would obviously present a problem when it came to getting the crew and ourselves into town for the Stampede.

My industrious father walked his TD-9 tractor to Churn Creek in the hopes of pulling some logs across to make a temporary bridge. To do this he first threw a rope across the torrent to a waiting Cattermole Trethewy logger on the Gang Ranch side of the surging creek. This logger grabbed and secured the rope, and Dad attached his TD-9 winch line to his end of the rope. Next, the logger pulled the winch line across and attached it to a couple of long logs that were waiting to be bundled up with cables and put into the Fraser River, which was how Cattermole shipped their logs to a mill at Agassiz.

At that point, our daughter Lisa was three and Baby Jack was only five months old. We would all have to make it across that little log footbridge if we were to make it to the Stampede. Churn Creek was a wild torrent and Liz was not at all sure that we should try this very scary crossing fearing—rightly so—that any misstep would end in disaster.

Jim Ledrew, a friend from high school days, who was helping out on the ranch, drove with us to the makeshift bridge. Liz watched in horror while I passed our baby to Jim. I meant for Jim just to hold Jack while I carried our three year old Lisa across, intending to come back for Jack. Halfway across the sixty foot log, I almost fell when I glanced back over my shoulder to see Jim coming behind me, carrying Baby Jack. Jim, for his part, had put on his best go to town clothes, complete with hard leather-soled Oxfords.

Understand that Churn Creek, below us, was an angry swollen torrent, and a mere fifty yards downstream is where it fed into the raging Fraser River. A fall would likely mean that the bodies—yes *bodies*—would be swept away perhaps for miles. This explained why Liz was so scared.

Despite the lump in my throat and Liz almost fainting from fright, we all got across safely.

Liz was never quite sure what I would do to present her with her next challenge. I know that she felt quite isolated in our little home on the range anytime I headed out to the mountains with cattle. She decided that she would take the children and go to Williams Lake to stay with her parents whenever I was away for any length of time.

When she moved to the ranch, Liz's dad, Doug Stevenson, had given her an old fashioned battery radio so she could tune in to CBC to keep her company. I subsequently borrowed the radio when she was in Williams Lake and I was in the mountains, so that I could listen to the news at night around the campfire. I later traded that radio to the redheaded girlfriend of Gordon Gerrard—one of my Gold Bridge cowboys—in order to buy Mike, my tracking dog. I did this, furthermore, without as much as a "by-your-leave" to Liz, who was furious at losing her only contact with the outside world. I assured her that I would get her a new transistor radio—the latest thing—and that Mike would repay us one-hundred fold. Mike certainly lived up to his part of the deal but not until the loss of Liz's radio almost lost me my head. The only problem with the replacement transistor radio was that it could not bring in CBC—a serious deficit.

Driving on the roads from Empire Valley to Williams Lake and back was an adventure because we were never quite sure what conditions we would meet. We drove the wintry roads in our little Volkswagen Bug. The Bug could handle almost any kind of winter conditions, what with it's engine weight over, and ice chains on, the rear driving wheels.

One day we were on our way back to the ranch from Williams Lake and we were halfway up our road from Churn Creek to the ranch when we started running into piles of drifted snow on the inside area of the road. I kept a real good head of speed on the outside edge of the road to negotiate around and partially through most of the drifts, but suddenly we were faced with no road at all, only drifted snow from the four foot bank sloped diagonally all the way across the road to the down slope.

I made an instant decision, with no warning to Liz, I turned the little car sharply off the road and down a ridge heading straight for the Fraser River a half mile below us. When I could swing across a little gulley to get to the next ridge I did so, again, with no warning to Liz. I then powered the car full bore up that ridge and jumped back onto the road just beyond the big drift. The white drifted snow was no whiter than my wife's face while she clutched her babies to her breast. (These were the days before car seats and seat belts.) It took her a while to get enough breath to let me know that she was terrified. Making instant decisions has never been a problem for me, but over the many years of our marriage, those decisions have caused Liz to have occasion to wonder how she managed to get mixed up with this harum-scarum Chilcotin cowboy!

I have a final stinky little story to tell you. The first winter we spent in our little poorly insulated frame house, I got the bright idea to build a little two foot wide enclosure around the bottom of the house. I then packed fresh manure from the barn's manure pile to fill up this empty space. I thought I might kill two birds with one stone. The fresh manure would give off heat while composting and prevent the cold air from creeping in underneath the house; in the springtime the manure would be a great base for Liz to grow a flower garden. The manure did a great job of insulating the foundation of the house during the winter, but in the spring, Liz started asking, "What's that awful smell?"

Those awful smells got worse until I covered the manure with dirt. That helped a quite a bit I thought, and while Liz was not in complete agreement, she did grow some beautiful, sweet smelling flowers. My wife has a <u>little different slant</u> on this story which follows:

I had never lived on a ranch. As a matter of fact, I had never wanted to live on a ranch, but when you fall in love with a handsome cowboy, what's a girl to do? After our July wedding, Mack and I moved into a little house on a hill overlooking the ranch. The house was small and poorly insulated. I had only lived in houses with central heating, and as the autumn winds blew coldly across our floors I wondered what the winter would be like.

"No problem," replied Mack. "I will build boxes around the bottom of the house, it will keep us nice and warm and in the spring we can plant flowers."

While the banging and pounding was going on outside, I was in the kitchen trying to learn how to cook. Sometime later I became aware that the very strong smell of manure was permeating our house. "Are there cattle corralled around the house?" I asked Mack.

"No," he replied. "I have just filled the flower boxes with manure."

I was in the early stages of pregnancy and was feeling ill. "How am I going to manage with this terrible smell?" I asked him.

"Don't worry," he assured me. "In a few days the manure will freeze, there will be no smell, and in the spring we can have beautiful flowers."

As things turned out, the manure did freeze and the smell did disappear. However, as luck would have it, that winter was one of the warmest on record. Every time the icicles began to drip, I knew instantly because our honeymoon cottage would be assaulted once more with the overwhelming stench of manure. I didn't realize it then, but that incident was a window into what my future life was going to be like, married to my handsome cowboy.

While I'm here, I could mention another incident that comes to mind when I thought it would be a good thing to take Liz and three year old Lisa out into the mountains to show them the beauty of the country I herded cattle into every summer. Here is Liz's version of what happened during that adventure.

As Lisa and I prepared for our exciting journey, I was sad at the thought of leaving Baby Jack, aged six months. Cappy (Mack's mother) had offered to look after him, but as I'd never been away from him before, I did feel anxious. I packed up his little clothes, his baby seat, and his special toy that he loved to kick. He would scream in delight when the jingle of a bell told him that he had once more made contact with the toy. We were proud of Jack's physical prowess—he had rolled over at five weeks of age, and he could hang onto his dad's fingers and be lifted into the air when he was just a little older. We were not to know until much later that he would become a superb athlete.

However, back to our trip. We drove from the ranch, through Lillooet, up to Gold Bridge then on to the Jewell Creek Bridge. There we met some of the cowboys with the horses we would use. I was riding with Lisa on the front of my saddle. During the ride, Mack and the cowboys were talking and laughing about their last trip over the shale bank that we were about to cross. Apparently, one of the pack horses had fallen off the trail, down 150 feet to the rushing

Gun Creek below, to be rescued by the cowboys, further on down the creek.

I was already worried, and hearing this tale didn't help. When we reached the trail I realized that, for 200 yards, our horse would have to pick his way on a very narrow path above that rushing water. As we started on that scary trail, shale was falling down onto the trail and on into the creek far below. How could I ever save my little girl if my horse fell off the trail and ended up in that churning water, I wondered? By the time we were on the other side of the trail, my nerves were shredded. I thanked God that we had made it safely. Luckily the rest of the ride was quite uneventful.

Liz and Lisa on Gun Creek Range

At the end of our journey, Mack set up the tent where he and I and Lisa would sleep and after a cowboy dinner cooked on the campfire, we settled in. The next morning, after a delicious pancake breakfast, Mack was off to tend to the cattle.

"What will happen about dinner?" I asked.

"Oh, don't worry," he answered, "Walter will cook the dinner."

I set about doing my camping chores, doing the breakfast dishes and tidying up the tent. Then Lisa and I took our clothes from the day before down to the creek and washed them. As a matter of fact, I washed our clothes every day in that creek and hung them on the bushes. When dinner time approached, Walter asked me if I knew how to cook grouse.

"I know how to cook grouse at home in my electric frying pan," I answered, "But I don't know how to cook it over a camp fire."

Walter claimed that he did not know how, either, so it was up to me to cook the dinner. I had my work cut out for me, trying for the first time in my life, to make a meal on a campfire. Lisa, of course, was delighted with the whole procedure and I had my hands full trying stay out of the smoke and to make sure she didn't get burned in the fire. When Mack arrived I let him take over the cooking. He told me that, of course, Walter had cooked many grouse over many campfires, but he was embarrassed to do it in front of me.

So, for the next few days I kept to my schedule of washing clothes in the creek and giving Lisa a bath in a little tub every night. The cowboys were having a good time chuckling at this silly woman. The child and the clothes were just going to get dirty again as soon as Lisa left the tent in the morning and fell into the dry, dusty soil.

Shortly after our return, Lisa, Jack, and I went to visit my friend, Gwen Oaksmith, in Seattle. Gwen and I had gone to UBC and trained as teachers together. Gwen was from Fort Saint John in the Peace River district of BC, and she and I, along with her friends from the North, Anita and Dolores, all became fast friends. Gwen had invited me to come to visit her in Seattle where she and her husband, Mauri were living.

The next time Mack was due to go back into the mountains he drove us down to the border crossing at Sumas where we were met by Mauri, Gwen, and their baby, Ellyn. The children and I stayed in their lovely home for ten days and had a wonderful visit. One night we asked Mauri's teenaged sister, Cammy, to babysit while we adults went out. At the end of the evening, when Mauri was driving Cammy home, she said to him, "That poor woman! That poor, poor woman!"

When Mauri asked why I was to be pitied she replied, "She has to live in a tent, wash in a creek, cook over a campfire—how does she manage it with two little kids?"

As it turns out, Lisa had been entertaining Cammy with tales from our camping trip. Mauri was happy to explain that, while we were isolated, we certainly had electricity, stoves and washing machines on the ranch.

While I have the floor, so to speak, I should tell you about the next trip I took to the mountains. A few months ago I happened to find the copy of a letter I had written to my dear friend, Myrna Evel, who at that time lived with her husband Jimmy and girls Catharine, Rebecca and Susan, in Burlington, Ontario. Myrna and I lived together when I had spent a year with her and her family in Hamilton. Her brother Rad was my boyfriend at the time. Myrna and I have maintained a close friendship over the years and have visited with our families at each other's homes. Mack and I have also kept in touch with Rad because he was a friend of Mack's, as well. In this letter I was telling Myrna all about the latest trip to the mountains. The letter was written in 1978. We were living in Kamloops at the time. I was teaching Grade One at Dallas Elementary School. Lisa was 17, Jack, 14 and Doug, our youngest, was 12.

Mack, Lisa, Liz, Doug, Jack; Liz completes 5th Year Education Degree, UBC, 1993, wearing her mother's graduation gown circa 1927, UBC

August 24, 1978.

Dear Myrna,

We were so pleased to see you and your family and it was nice to receive your letter to find that you had all enjoyed your stay. It means a lot to see you and Jimmy and the children. I'm glad our children are able to know each other—I guess it's our turn next time.

Our holiday this year consisted mainly of a trip to the mountains—on horseback. I thought a lot about you while we were gone, Myrna, remembering your fear of horses. I am not exactly a horse expert and I do not care for travelling in the mountains on horseback. However, the whole family was going and I was elected to accompany them. As the plans for the trip progressed, it appeared that I would have to drive our blue four-wheel drive, pulling a horse trailer.

The night before our trip we went to a party. I had wanted to get to bed early as I knew that we had a big trip the next day—and I had to drive that truck. You may not have a hard time understanding that Mack was having so much fun at the party that we didn't arrive home until 2:00 AM.

Starting the trip tired was not my idea of being prepared but off we went, later than expected. It took us a long time to pack up all the supplies and to get everyone organized. I was driving with Jack and Lisa, ages 14 and 17. I was trying to keep up with Mack on the highway between Kamloops and Cache Creek. While rounding a curve I felt as though I had lost control of the steering on the truck. Jack looked back and said, "Mum, you'd better slow down, the trailer is about to jackknife." Needless to say, my heart filled with terror. I pulled over, we got ourselves settled, and I decided to choose my own speed. It turned out to be only the beginning. After Cache Creek we had a six hour, torturous drive on a progressively bad road. We ended up spending the last three hours driving on a one-lane road from Lillooet to Bralorne. Because logging trucks were also on that road, it made life rather uncomfortable. Logging trucks don't have much concern for anyone smaller than they are, and, as they tear past you, they leave so much dust in their wake that it makes visibility almost impossible, which is somewhat terrifying on a curving mountain road. The fact that we were driving west at sunset didn't help either because the sun was in my eyes, the dust was boiling around us, and I had my precious children in the truck

However, we did manage to arrive at our campsite that night, with no more undue incidents, except for the fact that I was so exhausted that I couldn't get out of the truck. I lay down on the front seat while the children tumbled out to help Mack unload the horses. The next thing I knew, our horses were being challenged by the horses that already belonged on that property. With the sounds of neighing and kicking resounding in the air, I could hear Mack shouting to the children, "Watch out, you'll be kicked if you stand there." "Look out! Get out of the way!" All this was accompanied by the sounds of the horses kicking the sides of the trailer while the wild whinnying was renting the air. When the shouting and noise died down, Lisa came to the truck and said, "Mum, aren't you going to come and sleep in the tent?" "No," I replied, "I'm going to sleep here." "Aren't you even going to brush your teeth?" asked Lisa. She knew things were bad when her mother wasn't going to brush her teeth. Mack brought me a blanket and a pillow and I stayed in the truck for the rest of the night.

The big stock truck that Mack had been driving had carried five horses; my sister Jean had come up from the Coast and she had three ponies in her horse trailer. Her two little boys, Jody aged 8 and Barry, aged 6, were thrilled to be going on the trip.

The next morning when I awakened I felt much better, especially after I had brushed my teeth. Now I'd only be riding a horse and not have to worry about being forced over a cliff by those logging trucks. We all got saddled up and Mack put our supplies on the two pack horses. Off we went, down the trail, looking like the Raggle-Taggle Gypsies-O. We had four dogs with us, one of which was Jean's little poodle, who had to ride on the pony with Jean when the going got tough—which was most of the time.

A few yards down the trail we suddenly had to turn off and climb a bank. Once up the bank we were into heavy undergrowth, with dead trees that the horses had to pick their way through. The branches were hitting us in the face and I was sure one of the horses would trip and break a leg on all of the dead trees. Luckily that part of the trip ended before too long and we were once again on a proper trail. It led up a mountain and soon we were in beautiful alpine country. The wild flowers were magnificent and the grass was lush and thick. We rode up and up until mid- afternoon when we stopped for lunch in an alpine meadow. We were all resting after lunch when Jean noticed that two of her ponies were continuing up the mountain on their own. While she was getting them, the third pony, unnoticed by anyone, had taken off down the hill. By the time we realized what had happened, Mack was afraid that the pony would have returned to where we had left the trucks. He and Jack tore down the hill after the pony, caught him, and on the way back up the hill with the recalcitrant pony Mack noticed that he had missed a turnoff on the trail that we should have taken, a mile down the hill. The trail hadn't been used for a long time and it was very overgrown. So back we went and got on the right trail. The flies were very bad and they were really bothering us and the horses. My horse's skin was very sensitive and she seemed to blame me for the flies. Every so often she would reach back and bite my leg. If I had to get off for something, she would push me with her head, partly because she was trying to scratch herself, but I'm sure that part of it was strictly malevolent. I had Leslie Kalyk's horse, Josie. They had given me this horse because she had a lovely temperament and was the gentlest horse we had. Therefore, my complaints were ignored, because, after all, I had the nicest horse. As it turned out, Lisa took pity on me and rode Josie on the way back. She had almost as much trouble as I did. Josie, it turns out, is a lovely gentle horse at the ranch, but she was not pleased with the ride in the mountains. She blamed me for the trouble she was having and she was bent on punishing me.

We made camp that night beside a glacier where the flies were less bad. Our dinner that night tasted marvellous—even plain boiled potatoes take on a gourmet flavour in the wilderness. I crawled into my sleeping bag and dreamed of bears. As it turned out, we were not attacked in the night, and we all awoke feeling refreshed. It takes some time to get packed, as each pack box, two for each horse, has to be evenly weighted. Then everything has to be very thoroughly tied. Mack was the only one with the knowledge to be able to pack the horses but Lisa and Jack became pretty good at it after helping their dad for a while.

It was a beautiful morning, the sun was shining and the flowers were glowing with colour. Lisa was identifying them all for us and I was just beginning to feel really glad that I had come when Mack announced that we had lost the trail again. He knew where we should be going and, unfortunately, it meant going over the top of several mountains to find the trail. As we wound our way up, the grass and the alpine flowers gave way to large rocks and rocky gravel. I couldn't believe that we were actually going to continue. Every so often Mack would stop all of us and ride up the trail ahead to see what the terrain was like. To stop was a nightmare for me because my horse did not like to stop. When she had to stop she turned around and around and got herself sideways on the hill. If I got off it was even worse, as then she pushed me with her head and I was terrified that I would fall over the cliff and off the mountain. Finally we rounded one mountain top and to my horror, on the other side was a glacier that we had to slide down. Mack went along tightening everyone's cinches before the descent, and when he got to me I said that I couldn't possibly ride down the glacier and I felt that the children would all be killed. I was the only one howling; the rest of them were happily sliding down the glacier. Mack's pack horse, Duke, had started the parade. Duke felt confident as he was used to mountain travel. He sat down on the glacier and slid to the bottom as though he was on a toboggan. The rest of the horses followed suit. Mack led my horse down the glacier and I stumbled on down behind, wiping my eyes and feeling glad that I had put my hiking boots on that morning. One of Jean's ponies wandered over to the edge of the glacier, looked down over the cliff, lay down, and prepared to die. Jean tried to get her up but she refused to budge—she knew she was about to die. (I knew how she felt!) Mack went over and gave her a kick with his boot and yelled at her. She forgot about her death wish, got up and joined the others. When we got to the bottom Mack *looked up jubilantly and said, "There's our trail!" Jean and I looked up at what appeared to be a pencil line drawn in bare rock on the next mountain. We expressed our disbelief. However, it turned out to be wide enough for the horses and, with the flies still madly biting us, we continued on our way. We finally arrived at Spruce Lake, where we were going to set up camp and stay for a couple of days. It looked marvellous to me. The only cloud on the horizon was that fact that I knew we had to go back over that same trail.*

While we were at Spruce Lake I was sure that nothing traumatic could happen to us. I arose the next morning, full of pleasure at the thought of spending all day off a horse. Jean, Lisa and I took our clothes down to the lake and had a great time washing them. That is, we did until the flies found us and were delighted by the fact that we were wearing our bathing suits. We fled the lake with our clothes less than clean and hurried to cover up all that bare skin. The flies at Spruce Lake weren't nearly as bad as the ones we had encountered during our ride, but it was a good idea not to leave too much skin free to tempt those that were there.

This brings to mind another difficulty about life in the wilderness. Whenever one had to take a "nature break," it was with great unwillingness that we would expose our tender nether regions to the onslaught of those damnable flies. We girls, of course, were in much the worst position, but I guess that is something that even Women's Lib will never be able to do anything about.

When travelling in the mountains, one by necessity must travel light. Therefore, when my period wasn't due for ten days, I had taken only the most minimal of supplies. However, you guessed it, what with all the bouncing and shaking around that I got, I also got my most unwelcome little friend. What do you do when you're miles from nowhere? Why, you rip up towels and you sit –quietly, as that's all you can do, with those towels flapping around. What did women do before modern techniques came to their aid? I will never again take such simple articles for granted and I will never again go unprepared into the wilderness, even if I have to take an extra packhorse with me,

The next crisis came when Doug became ill with a fever. When you are so isolated it is a worry to know that it would be a major effort to get to a doctor. However, Doug survived, and I survived, everyone caught lots of fish and by the time we got home I even found myself promising Doug that I would go again next year to make up for the fact that he got sick. Want to come with us?

THINGS NOT TO DO WITH YOUR CHILDREN

Everybody would like to do some things over again or live a part of their life differently. I am no exception. I offer a couple of reasons to mend your ways, or should I say, my ways. One fall day when Lisa was four, I was helping the threshing crew harvest a big crop of oats in fields below the main buildings.

Here's how it works: A binder cuts the grain, rolls it into a bundle and drops it on the ground. Four or five bundles are gathered up by hand and leaned together forming a stook. After they dry, these stooks are taken to a threshing machine that then shakes and separates the grain kernels from the straw.

My job was to load a wagon with grain stooks and pull the outfit up out of the fields, across a raised bridge, which crossed an irrigation ditch to the threshing machine. Dad had set up the threshing machine next to the barn so we could unload the grain into the barn loft.

One of my worst problems was that I was always highballing to get the job done in good weather. I had loaded my wagon by pitchforking the bundles over my head to a height of about ten feet off the ground and onto the wagon. Lisa was with me then, as she usually was when I was working close in at the ranch. Once the wagon was loaded, I chucked Lisa up on top where she could have a nice soft ride to the threshing machine. I told her to "Hang on!" forgetting that this was her first time to ride on top of a load. I was used to the Gillis children who were old hands at knowing what to do on ranch equipment.

All went well till I hit the bridge in fourth gear and was travelling about fifteen miles an hour with the big Massey tractor. Unfortunately for both Lisa and me, the wagon wheels bounced and shook the load severely as we drove over the raised bridge. As a result, Lisa tumbled off, or should I say she was thrown off, face first, into the dirt. Her poor little teeth went through her lip and there was blood everywhere. I picked Lisa up and ran for my house about three hundred yards away to give Lisa to her mother. Liz was quite shocked to see her little girl covered in blood, and not too pleased with me, needless to say, but she quickly got everything under control.

To this day Lisa has a great fear of heights that no amount of persuasion can reduce. After that our outings were somewhat curtailed, but Lisa did bounce back and some of my fondest memories are of taking her out fencing. She was my constant companion in the springtime as we would re-build and repair the hundred miles or so of fence at Empire Valley.

While I shovelled new posts into holes and repaired broken barbed wire, Lisa would entertain me with stories or have her own little adventures. She learned an awful lot about vegetation, animals, and insects that she never forgot. One time she discovered a colony of yellow ants and of course we had to take them home—where they got loose in the house. That episode didn't make Liz too happy.

Another time she captured a few garter snakes and took them home in our lunch box. That was fine until she lost one in the bathroom of the Gillis house. Don, as I have already related, was deathly afraid of snakes—evidently a phobia which cannot be erased. So when Lisa arrived crying at the cook house poor Don was none too pleased to hear, "Uncle Donald, can you help me? I lost my snake in your house."

There were some pretty harsh words from Lisa's uncle. "Lisa, you find that snake and get it out of my house, or you'll have to come and sleep in my house tonight and I'll sleep in your bed."

I'm not sure how those threats turned out but I know Lisa never found the snake. Donald would not have fit into Lisa's bed, anyhow.

Jack, for his part, got me into some really big trouble when he was about two years old. He had headed down to the big house from our little house, telling his mother he was going to visit Cappy (my mother.) As he passed the Gillis house he stopped to watch John, who was about twelve at the time, cutting the lawn. John stopped his lawn mower to pick up an old BB gun out of the long grass in front of the mower. "I'm going to shoot you, Jack Bryson," he said as he pointed the Red Ryder BB gun at Jack.

Jack didn't say a word but spun on his heels and headed back for our house. John felt a little badly that he had scared the child but positions were reversed a few minutes later when Jack arrived back carrying my .270 rifle. With difficulty he hoisted it up, pointed the barrel at John and said, "I'm going to shoot you, John Gillis."

"Jackie, Jackie, give me the gun," a white-faced John Gillis pleaded.

A few minutes later Liz was surprised to see John pack Jack and the .270 through our front door. Liz was vacuuming, and so she had not heard Jack come back into the house, go into our closet, and disappear with the rifle. You can bet your bottom dollar that the guy that got chewed out the worst wasn't John or Jack. I still haven't heard the end of that one.

While we are talking about kids—my favourite story about my daughter Lisa was a direct result of her always being around her dad on the ranch. She was endlessly inquisitive, like all kids, but I am not sure that all kids talked a mile a minute as Lisa did, or asked half as many questions.

Lisa really liked to help me when I was helping out the heifers in the calving pen across from our house. We usually calved out three or four hundred first calf heifers in that large pen. Anybody who thinks that isn't a dynamite chore hasn't calved out a lot of heifers.

We kept a pretty good watch over this young stock that was intended to calve out in the month of March when they were only two years old. We hoped the snow would be gone by then and that there would be some warming sun to dry off the newborn calves as their young mothers weren't always that careful about licking them dry.

There were always a lot of problems associated with the calving process which required someone to assist about twenty-five percent of the mothers with their newborns. In retrospect, I should not have bred the heifers to calve as two year-olds but rather bred them to calve as three year-olds, when they would have been bigger and much better prepared to experience the birthing process.

Normally, with these troubled two year-olds, we would drive the straining mother into a catch pen and leave her for a few hours to see if she could consummate the process on her own. If they could do this on their own, these heifers would then bond with their calves and get them up and sucking a lot better. When they couldn't, we would usually resort to hobbling their front feet so they couldn't move around much and then go to work with a calf puller and an obstetrical chain.

Because we calved our heifers out as two year-olds we consequently had a lot more trouble than those ranchers who calved out first calf heifers as three year-olds. My thinking was that in the scheme of things we could get one more calf in the calving life of a cow if she were bred as a long yearling and calved her first calf as a two year-old.

Lisa picked up a lot of tricks from her dad to help the heifers kick out a calf, help to keep it alive and finally get up on it's feet to suck that first colostrum milk, which is so vital to a newborn. Just using a calf puller properly without damaging the heifer or the newborn is a work of art. Lisa learned a lot of tricks like rubbing snow on the forehead of the calf when it didn't immediately start breathing. If there was no snow, then tickling the inside of the calf's nostril with a straw would help to start it breathing. We even gave mouth to mouth resuscitation and heart massage as last resorts. Lisa got a chance to demonstrate her prowess at calving when Liz took her to Williams Lake to attend kindergarten, at age five.

Liz and the children stayed in a cottage on the Stevenson property, a few hundred yards from her parents' house. In the spring Lisa's kindergarten class in Williams Lake took a field trip up the highway to a Soda Creek Ranch, about twenty miles out of Williams Lake, to see the rancher's new baby calves. When they arrived, the rancher was having a tough time. He was lying on his back, trying to wrestle a calf out of a heifer while using his calf puller. Since he obviously didn't have time to show the kindergarten class and their chaperoning mothers around, the kids clambered up onto the top rail of the corral to observe a real live birth. The problem was that the rancher was having such great difficulty that it was obvious to Lisa, perched on the fence, that he needed help.

"Mister, Mister, you're not doing it right," volunteered Lisa. Something must have triggered in the exasperated rancher's mind because he stopped fighting the heifer long enough to ask Lisa who she was and how come she knew that he was not doing it right.

"My daddy doesn't do it that way," she offered and clambered down off the fence to demonstrate. "You have to pull the calf puller down in the direction of the cow's heels, not straight out and only use the jack at the same time as the cow strains," explained Lisa.

Following these expert instructions, it wasn't long before the new calf came out of his constricting prison and started gasping for breath. The rancher was ecstatic that this little five year old had shown him how to use his calf puller properly and I'm sure he used that knowledge many times over in the years thereafter. Lisa, of course, was the hero of the day for the class. Her teacher and the assembled mothers were probably the most impressed witnesses to this remarkable little girl's feat of veterinary medicine. Strangely enough, she didn't become a vet or a doctor. It was Jack "I'm going to shoot you, John Gillis" Bryson who became the doctor and practices on the Sunshine Coast. Lisa is a social worker, helping people, not cows, in Vancouver.

Jack packing up at Spruce Lake camp—in the 1990's

Chapter Eight
Beans Up—Beans Down

When you pack continuously into the backcountry as we did at Empire, you cannot always gauge exactly how much food a crew will eat over a two-week period. Two weeks was probably the average number of days it would take us to drive one herd of cattle into the mountains in late spring or to bring them back out again to the fall and winter ranges. We always erred on the side of too much food but even then you can run out. Nothing gets you down in the mountains like being out of food. A lot of my stories are about that situation. I have talked about Don Tremblay and me finding grouse and wine on Poison Mountain on Thanksgiving Day. That find saved our bacon, literally.

On one cattle drive to the Tyax country, I had my sister Donna and her husband Don Gillis with me, plus my buddy from UBC, Nick Kalyk with our usual cowboys, Walter Grinder and Francis Haller. In addition we had John Moss from the Pavilion General Store, which at that time was the oldest standing General Store in British Columbia —previously owned by my uncle, Ernie Carson. Unfortunately it burned to the ground in the year 2000. John brought his wife Bubby, who was sister to my Uncle Duffy's wife, Dorothy. Duffy and Dorothy had their own ranch at West Pavilion.

All went well on that particular drive except that we could not cross the Tyax River ford to let us take the cattle into the Spruce Lake/Gun Creek country because of raging high water levels. We tried, but some cows were swept downstream, as well as Don's dog, Jack, who got out by himself a mile downstream. So when the dog returned and the cows were safely out of the river, we pulled back and left the herd at Bannock Camp in the Tyax River country, under the care of Francis Haller.

Because Francis would be there most of the summer, waiting for the river to go down enough that he could safely shove small bunches of cattle across the Tyax ford, we had left him most of our remaining grub. We had obviously shortchanged ourselves a little on the food menu, for by the time our crew hit camp at Starvation Canyon—aptly named it would seem—on the way home, we were out of grub, except for flour.

Now some packers or guides had left an old camp at Starvation Canyon, which had much deteriorated over the years. Snow had flattened out the old tent, and it had rotted away. All the gear was strewn about. I was aware of the old campsite and thought there just might be some canned food or something edible in some old rubber pack-bags left at the site.

Sure enough, some scrounging on my part produced a half can of old peanut butter. We spent a pretty hungry night at Starvation Canyon and in the morning I determined to put the two items of food that were left into use. I mashed the peanut butter into our remaining flour and added a little water, thinking it might make pretty good hotcakes over the coals of our fire. Everyone partook, for they were semi-starved.

"I'm so hungry," Don Gillis said, "my stomach thinks my throat's cut."

Those hotcakes didn't taste at all good and they stuck around in our bellies for at least the next forty-eight hours. To this day I cannot eat peanut butter— it is the only food I can't abide. There must be some genetic connection to this event, for my grandson, Malcolm Bryson, son of Jack and Carla, is deathly allergic to peanuts. With my history, I am not surprised. I am sure none of us could look a hotcake in the face for at least a year after that starvation diet in Starvation Canyon. (As an aside, years later when we were visiting Len and Donna Marchand in Kamloops I remarked to Donna, "These cookies are so good, Donna, what do you call them?" "Peanut butter cookies," replied Donna with a twinkle in her eyes.)

The title of this chapter promised you beans though, and beans are what I mean to serve up—as I said, it is hard to be certain how much food to pack but also it was difficult to keep meat fresh for very long. Nights were usually cold but days were often hot in spring, summer and fall. The answer to spoiled meat was to pack canned meat: Prem, Spam, ham, canned bacon, and lots of beans.

Wonderful golden brown pork and beans—the staff of life— ambrosia to hungry cowboys like me! I love them and always packed a case of pork and beans on our trips. Unfortunately, not everyone enjoyed them as much as I did. Not so much because of their backfiring, but because some folks just plain don't like beans. One of those was Louie Seymour from Canoe Creek, our old one-eyed Native cowboy.

Louie had elected to stay on with us after Henry Koster left. Even though he had only one eye— the result of hitting a dead tree limb at full gallop—Louie was the only one left at the ranch who really knew the ranch proper, when we took over. Consequently, he was indispensable if somewhat cantankerous.

He was a bit hard on horses' backs with his old saddle, but he knew the country, and he knew the cattle. He had a great deal of difficulty getting up onto his horse's back on that old saddle but he wouldn't hear of trading it for one of my newer ones. I guess though, when you are as old as he was, you get a little set in your ways.

The old boy was pretty ancient and he had great difficulty getting his old bones onto his favourite ranch horse, Rocky. He would hook the horse's lead rope around the saddle horn and pull himself into the saddle that way. I estimated that he was about seventy-five years old, but when asked about his age, Louie didn't really know. "All I know is I was born late in the spring," was his stock answer.

What really made him cantankerous the most, though, was beans! He put up with my love of beans for one whole summer with only grunts of dissatisfaction, but when the cookhouse cook put canned green beans on the table in late fall, Louie had had it.

"Beans in the summer, beans in the fall, beans up, beans down. I Hate Beans," he growled as he left the cookhouse in disgust. It wasn't long after this that old Louie asked for his paycheck, and with that, he returned to his little spread near Canoe Creek, never to return.

As you've already read, I have pulled my fair share of tricks on various cowboys and wranglers in my time; it's only fair that the tables be turned once and a while.

One fall, I was headed back to search for strays in the backcountry after our main fall gather had been done. Les Kerr, from Langley Air Services, had been hired to fly his Super Cub into our backcountry searching for strays. We spotted a few between Red Mountain and Walter Grinder's Camp near Big Basin.

Now pretty well all our standby cowboys had left for the winter and so I was faced with heading out alone to bring in those last few strays. When Jimmy Joe, our ranch irrigator, who functioned as chore man during fall and winter at the ranch, heard that I was going by myself he volunteered one night at dinner in the cookhouse that he would be agreeable to helping me out.

Jimmy had been working for Dad since they were both young men at Pavilion on the Bryson Ranch there. He later worked for him when Dad was the manager of the Diamond S at Pavilion, and later still he worked at our ranch in Merritt. We are talking probably thirty years or more of uninterrupted slaving away for four dollars a day plus room and board. He was more of an old friend than a ranch hand.

The way he got on with Liz and the kids, he was almost part of the family. I remember one year, at the Williams Lake Stampede, Liz was staying in Williams Lake with her parents as I was heading back into the mountains. She had driven with me to the Stampede Grounds to pick up the crew so I could head back to Empire. Jimmy approached Liz to say, "He's leaving you on your anniversary."

Jimmy had never before spoken to Liz. He had never married and he couldn't read or write and so Liz was very moved to think that he knew our anniversary date and the significance of my leaving her on that special day. In fact, when we were first married, Liz had been uneasy around Jimmy because he was very quiet and kind of inscrutable the way some cowboys are. She made sure, after we left the ranch in '67, that Jimmy always had a Christmas card with pictures of the children.

Anyway, back to this particular fall trip into the back country looking for strays—Jimmy had taught me how to ride and also to hunt when we were on Pavilion Mountain and so I was happy to have him along for company, if for no other reason than that. Our first night out we stopped at Walter's Camp on the old trail that ran through China Lake, Fareless Creek, Yodel Camp and part way to Red Mountain Meadows. Walter had a nice little cabin there in the swamp meadows surrounding Black Dome and Fareless Creek. He used the cabin regularly in the fall to guide moose hunters. From this cabin he also hunted sheep and deer along Churn Creek and in Big Basin. I had stopped at Walter's Camp a few months previous, on my way home to Empire and had left our remaining cache of food in the cabin—including of course, that cowboy staple, canned pork and beans.

Fearing an approaching freeze-up, I had put the canned beans that were left over in the oven of the cook stove, hoping that the insulated oven might prevent the cans from freezing solid and bursting. Forgetting where I had put the beans, I asked Jimmy to get a big fire going in the cook stove while I unpacked. I put a couple of steaks on to fry for our much anticipated dinner.

It was later, after we had eaten the steaks and were sitting back on the bunks on our rolled out sleeping bags, picking the last of the meat out of our teeth with hand-carved toothpicks, that the cans of beans blew up—and I do mean *blew up*, as in exploded!

It took Jimmy and me a few minutes to realize what had blown the door off the stove and scared the pants off us, but gradually we saw the humour in it and after cleaning up the mess, we settled in again to ponder our next day's travels. Suddenly there was this god-awful howling sound outside the cabin door. The closest thing we could think of was a werewolf. Neither of us wanted to venture outside to find the source of those sounds, so we pushed Cuff, the cow dog, out the door and slammed it behind him. He immediately went into a frenzy of barking, and the next thing we knew the cabin door burst open. Jimmy and I both knew we were done for.

"Gotcha, didn't I?" exclaimed brother Dunc. He had voluntarily decided to bring up some extra provisions and bales of hay from the ranch. He had deliberately parked the four-wheeler out of earshot of the cabin and crept up to scare the bejesus out of us, which he certainly did, no question!

I still love beans, although exploding beans combined with werewolves could cause some folks to give up those cowboy mainstays.

Chapter Nine
The Gold Bridge Bar

As I have previously mentioned, the quaint, historic little town of Gold Bridge, located on the eastern edge of the Coast Range beside the Bridge River Highway, gave instant access to some of the world's finest scenery, including hundreds of miles of snow-capped peaks and lush green alpine meadows. These days it is also a kicking off point for folks wanting a backcountry trail-riding experience.

Not only was the country up there beautiful, but for me and the cowboy crew, the Bridge River Highway also gave access to Lillooet. From there it was over Pavillion Mountain to Clinton and on to Empire Valley. This route was our supply lifeline when it came to food, gear and grain for our camps on the Tyax and Gun Creeks. From the ranch, the trip took about six hours by four-wheel drive. It was long, tortuous, and somewhat nerve-wracking because of the dreaded hairpin turns and yawning chasms that fell away from the road between Lillooet and Gold Bridge. Nevertheless it was a lot better than packing supplies by horseback over the ninety miles of trails from the ranch to the Gun Creek cowboy camp.

To give one an idea of just how scary that Gold Bridge section of highway was to the uninitiated, I will tell you another Liz story.

On this occasion, we were about to take a cattle drive out from the home ranch via Spruce Lake to the Gun Creek Range. As I recall, there were probably a half dozen cowboys with all the gear necessary for a cattle drive of about four hundred heifers and twenty-five bulls. The normal sequence of events was slightly confounded by my UBC pal, Nick Kalyk, setting the date for his wedding to Margaret Borsato a mere sixteen days subsequent to the departure date of the drive. For me to make it to the wedding, I had to do some calculations.

I had estimated it would take approximately two weeks to get the cattle to the Gun Creek Range and one day for me to ride down to Gold Bridge and hopefully meet up with Liz. She was to drive our legendary family Volkswagen from Empire, up the Bridge River Highway to Tyax Lake and from there up an old mining road towards the Manitou Mine, situated at the Junction of Mud Creek, Relay Creek and the Tyax Rivers.

The idea was that we would run head on (not literally) into each other on the Manitou Mine portion of the road between Tyax Lake and the Manitou. Liz was game to go, which was to her credit for the trip would be daunting for even an experienced back roads driver. She, of course, had never driven so much as a single kilometre of the route.

Anyhow, she managed just fine until she met a logging truck going flat out with a load of logs on a dusty, rocky corner of the Lillooet/Gold Bridge section of the Bridge River Highway. The logger was going to need a good portion of Liz's lane to make the corner and, to complicate matters, there was a five hundred pound boulder resting precariously in the middle of what road was left to her. Faced with the split second choice between a fully loaded logging truck and a boulder, Liz chose the boulder.

The unitized body construction of the Volkswagen saved the day. The Bug hit the boulder and was airborne. Liz bounced high enough that she sustained a pretty good bump on the top of her head when she collided with the car's interior roof, but she did save herself from a collision with the high-balling logging truck driver.

She stopped to examine the car for damage and had a little prayer session before she continued on to the turnoff to Tyax Lake near Gold Bridge. Now she was really on a wilderness road, and again she did just fine until she eased around a sharp bend and almost ran headlong into a BC Forest Service truck coming at her on the single lane track.

The driver of the green BC Forestry truck seemed to think that no one else should have been on that road on this, his chosen day to drive it, so he took to chastising Liz for spoiling his day. After they got themselves around and past one another, Liz went through another prayer session before she continued on, feeling that her husband must be just around the next corner.

Now here's where fate played a hand, for unbeknownst to either of us, Liz had already passed me by, for I had gotten an early start out of our camp on Relay Creek and ridden the twelve miles to Tyax Lake before Liz arrived. Seeing a female resident outside her cabin on the lake I figured that this person might put up my buckskin horse, Buckles, for a few days, while Liz and I went off to Nick and Marg's wedding.

Just at the very time I rode the 200 yards down to the lake, Liz had arrived at our prearranged meeting place. She had waited ten or fifteen minutes before she decided to drive on to meet me—surely I must be just ahead on the road.

A COWBOY'S LIFE

For my part, I did not hear Liz go by, for I was probably in earnest conversation with the cabin owner. After a time, Liz returned down the road from the Manitou, having met a hiker who told her that I had passed him going South earlier that morning. Needless to say, she was in a bit of a state after narrowly missing being killed by a logging truck, scolded in a confrontation with a tactless and ungentlemanly Forest Service officer and missing me, her dilettante husband. She had an additional fifteen miles of terrible back road driving, trying to find me. She returned and took the time to point out, in no uncertain terms, that had I been on the road, where I should have been, we would have rendezvoused within minutes of our appointed time set two weeks previously at the ranch headquarters on the Fraser. I think Liz, to this day, is still irked by the fact that, after all her tribulations, she was on time at the right place and would have picked me up then and there, had I not wandered off.

But back to Gold Bridge bar—Back in the summer of '58 when Don Gillis, Walter Grinder and I had first followed that old BC Resource map from Empire to Spruce Lake, looking for additional cattle ranges, we eventually decided to head for a little spot marked on the map as Minto City, sitting aside the Bridge River about six miles south of Gold Bridge. We figured to provision up and spend the day in the local hostelry as we were sure that there would be a bar in a place that was referred to as a "City."

Fifteen miles of dusty trail did indeed bring us to Minto City. Imagine our chagrin to find out that it was not a city, but a ragged community of approximately sixty people with no facilities whatsoever. Furthermore, it was soon to be drowned by the rising waters behind the Carpenter Lake Dam—today BC Hydro's main power supply for Vancouver. The water is piped through the mountains to Shalath. The power generating turbines in Shalath then send power to Vancouver via high voltage power lines. The despondent residents in 1958, who were busy packing up their household belongings, pointed us in the direction of Gold Bridge if we wanted supplies and a beer. So we continued on, by now a bit saddle weary and very thirsty from the long trek out of Spruce Lake's cowboy camp.

We were not disappointed by Gold Bridge, as it had everything a cowboy could want—a horse hitching rack out front of an old western style hotel, with a saloon/bar. In true western tradition, we tied up our broncs at the hitching rack and renewed our supplies at the general store. The picturesque old hotel had six rooms complete, of course, with the obligatory bar and owners who were willing to stake us to a meal and a room on credit, using our ranch name as guarantor.

The owners had heard about the Empire Ranch out here in the goldmine country and typical of a country hotel, they were willing to take a chance on us, something that probably would not have happened anywhere else and certainly would not happen today. Naturally the hotel bills were all honoured by my mother, the bookkeeper back at Empire.

After a wash up the three of us headed into the beer parlour, as they were called in those days, and ordered up a round. It is worth mentioning at this point that Empire cowboys never packed liquor of any kind on our cattle drives. We also kept very little at the ranch, because of the disastrous effect that booze can have on cowboys of all extractions. Also, our family was still under Eliza Jane Magee's decree against the imbibing of liquor. (It ended with my generation.)

We three caballeros, when confronted with a never ending spigot after a long dry spell, started belting down drinks. So it was, after the first half dozen beers had been quaffed and the dust in our throats had been washed away, that Don leaped to his feet and addressed the twenty or so miners who were enjoying a quiet beer after a long day underground, moiling for gold in the drifts of the Bralorne Mine.

"All right, miners against the cowboys."

This was said more in jest than in what might have been perceived as brazen confrontation to the boys at the bar.

Oh my God, I thought, as I quickly checked the exits. *We had ridden all this way for a beer only to die at the hands of twenty rock drillers on the Bridge River if they take this the wrong way.*

However, as luck would have it, Don was not only a good judge of beers and bars, but he was pretty astute at sizing up men. Before I could beat a hasty retreat out of the bar, we were surrounded by a half dozen garrulous miners, offering to buy us more beers in return for some good-natured storytelling and the possibility of buying fresh beef. One thing led to another and before long we were entered into contests of strength as more beer got Don's macho juices flowing as fast as the beer kept coming.

"I'll wrist twist anyone for the next round."

And before you knew it, he was vanquishing miners, first with his left hand, and then with his right. Of course, they wanted to twist me too; I looked quite a bit easier than Don, who was an imposing, muscular six feet two inches and 210 pounds, while I was an unimpressive six feet, and 150 pounds. My strategy was to let them all tire themselves out on Don's muscular arms with the promise that I would only wrist twist any miners who were able to defeat Don.

Of course, his arm eventually weakened and he turned them over to me. Bets were now for cash though we had none on us. Luckily, Don had already tamed most of the stronger miners, so it wasn't that difficult for a slender, wiry, trail-hardened cowpoke to pick up some good coin of the realm, before we retired for the night to our fancy—it seemed fancy to us—room in the Gold Bridge Hotel.

I still have many memories of Gold Bridge, including some from my uncle Duffy Bryson who had his first job off the family's ranch at Pavilion, on the road crew operating out of Gold Bridge in the thirties. He recounted tales to me of the "ladies" who plied their trade in the little shanties that still exist around the old hotel and the camaraderie offered by miners and road repair crews alike at Gold Bridge. The same kind of easy camaraderie we had received from those Bralorne miners even after Don's confrontational announcement, "All right—-miners against the cowboys," in the summer of '58.

Chapter Ten

Dogs That Have Trusted Me

CUFF

When we purchased Empire Valley from Henry Koster in 1956, we were fortunate to inherit a black and white Border Collie dog named Cuff who came along with the deal. Cuff was a Border Collie like no other. He wasn't much to look at; the expression "dog eared" comes to mind. Cuff was black and white all right but he had an unusual dish face and a disposition that was almost human. Because he was a ranch dog he really had no master, and consequently he could pick and choose amongst those who he worked with. He alternated between Henry Koster and his crew of Grinder brothers, Johnny, Henry and Walter, who all revered him and, by '56, had elevated Cuff to a rank all his own.

The Empire Valley crew knew Cuff belonged to the ranch and vice versa. I quickly came to realize the same thing. Cuff considered himself to be the reigning head dog and would, therefore, pay little attention to his scruffy comrades when he was working cattle. He particularly liked working with the Grinders because they treated him with the respect due a dog of his stature and, furthermore, they always got him home in time for dinner.

Most Border Collies work on hand signals or chirping noises but Cuff needed none of these. He had long ago mastered the intent of his drovers and so he was really his own boss. However, he would use the handiest cowboy's horse when a fight broke out amongst the *lesser* cattle dogs. In those instances Cuff would launch himself onto the back of a horse or an empty saddle and gaze disdainfully and dispassionately at the assorted rabble snapping and snarling below him.

My favourite tale about Cuff concerned my brother Dunc. We were entertaining our cousin, a nursing student named Heather Carson—daughter of Bob Carson, the realtor who had negotiated the sale of the ranch. Heather had brought half a dozen young, attractive female nursing student friends from Vancouver General Hospital to Empire for the Easter holidays. Dunc became quite enamoured of Sandy, one of the nursing students. This of course could never develop into any kind of long term relationship, for she was totally involved in graduating as a nurse. Also, Dunc was very involved with his job on the ranch where he was in charge of the farming operation, just as I was in charge of the cattle operation with my partner Don Gillis.

As it happened—Dunc wanted to take the girls for a ride on a beautiful Easter Sunday morning. So he and friend Jim LeDrew headed up into the overnight saddle horse pasture to pick up enough horses for the contemplated romantic ride. Jim came to the ranch almost every year to help Dunc keep our farm machinery running in good order. Jim operated his own heavy duty truck and he was a mechanic par excellence. Cuff chose to go along as nothing else was doing on the ranch that day. After an hour or so I became concerned that Dunc and Jim weren't back, for though the overnight pasture was big, it was not *that* big.

To my relief, a descending dust cloud coming down the steep side-hill from Grinder Creek indicated that the horses had finally been found and were on their way to the corrals. As I watched, the dust cloud became twenty or so horses that quickly disappeared around the barn and through the open corral gate. But no riders followed. Where were Dunc and Jim, and why were the horses there?

After a while our two hopeless romantics returned to explain that, to their consternation, they couldn't find the horses. The fence must be broken, they suggested, and I better get up there right away and fix it. I directed them to the corral where, to their surprise, the horses were standing with Cuff lying patiently waiting in the corral gate. The look on that dished face as much as said, "Where the hell have you guys been? I could have used some help."

MIKE & RIP

Mike was another Border Collie dog like Cuff, but he was younger, bigger, and more aggressive. He came to Empire through Gordon Girrard. Gordon and his family ran a guiding and packing business from their home on Gun Creek. Gordon was another of the young horsemen that Francis Haller had encouraged to seek employment at Empire. Gordon worked only a few months in that spring with Empire, before his red-headed girlfriend arrived with her dog Mike to take Gordon to the July 1st Williams Lake Stampede.

Maybe Gordon had long range romantic ideas for he came back from the Williams Lake Stampede only long enough to ask me if I wanted to buy Mike from his girlfriend. He had decided to move back to his dad's guiding business on Gun Creek near Gold Bridge. Gordon Girrard Senior was well known as a guide who pushed himself and his hunters very hard so they almost always got their California Bighorn Sheep.

As for Mike, I liked his spirit. Gordon and his girlfriend were wary of hanging on to the dog in the game country Northwest of Gold Bridge, because Mike had one failing: he loved to chase deer. After a little haggling, I traded my wife's battery operated "CBC" radio for Mike and so began our career together. You don't have to be a lawyer to understand the consequences of my trade for I did not ask for Liz's permission.

First off, I had to rid Mike of his love for chasing after the scent of deer. In Empire Valley this was not easy because deer were everywhere, but I persisted in watching Mike closely, and I verbally harassed him whenever he headed away from me with his nose on the ground at a dead run. Soon he would look back with a guilty look on his face, like a kid caught with his hand in the cookie jar. This guilty look began to happen now whenever he planned to take off after deer.

Gradually I weaned him off his predilection for deer scent and got him interested in working cattle with my other cow dog, Rip. I had purchased Rip for $100.00 from Rip Grey at the Williams Lake Cattle Auction. I could little afford the money for the pup, but the auctioneer at the cattle sale in Williams Lake had given the litter of pups Rip Grey offered a top billing so I threw away my second thoughts and ended up with the pick of the litter. Rip was a bitch and much smaller and more timid than Mike, but the two made a great team. After a while I had only to ride down the trail behind a herd of cattle using hand signals and chirping noises to have one dog work either side of me, bringing along the left and right flanks of any herd I was driving. Their habit was to tread softly along behind a reluctant cow and when she looked away they would nip the cow's heels to spook the animal forward. These dogs worked without ever barking and that is the secret to a successful cow dog. A cow with a calf will turn immediately, if a dog barks, in order to protect her calf. Barking dogs are all right on steers, heifers and dry cows, but they are a disaster around wet cows that are protective of their calves— as any rancher can tell you.

Mike's real value to me and the ranch came quickly to light the first fall I owned him. On one of our last fall forays into the Spruce Lake/Gun Creek area to pick up stragglers, I was by myself with Mike and Rip for company, looking for cows near Spruce Lake. Mike was ranging fifty yards ahead of me, with Rip trailing my horse. Suddenly I noticed Mike revert to his guilty look when he stopped to look back at me and I thought, *damn his hide, he's back to his old tricks.*

When I called him back and scolded him, Mike looked properly disconsolate and I reckoned that would be the end of it. Upon taking a closer look at the trail ahead I could see no deer tracks, only fresh cattle tracks—just what I was looking for. Mike seemed to have discerned the scent of cattle before I had seen the tracks. A feeling of relief came over me that Mike may not have been thinking "Deer" but instead may have substituted cattle scent in place of deer scent for his own adventurous purposes. I conveyed my own guilt and new expectations to Mike in a little one-on-one counselling session, which buoyed his ego considerably. Then I sent him off trailing the cattle with hand signals and chirping noises.

Away went Mike on the run with me on the trot after him. Every two or three minutes I'd call him back if he got out of sight. He quickly learned to come back and check on me if he got too far ahead. In this manner, after about half an hour, we came upon some cows lying hidden in a small spruce meadow.

It was halfway up the timbered side-hill above Spruce Lake in an area that I probably would not have searched for any stray cattle. Mike stopped as soon as he spotted the cows, with one foot raised in the air, staring straight at the cattle without turning his head. *By God*, I thought, h*e's like a well-trained bird dog. I've really got something here.*

Slowly I worked my way around the bedded cows in the timber while crooning a cowboy love song to let them know that I was there and meant no harm. When I eventually showed myself above them, they took off at a run down the timbered mountainside with Mike, Rip, and me in a controlled pursuit. I knew Mike would never lose them now, so we were able to get the cattle slowed down after a while and then herd them across the Tyax River on the first leg of their journey back to the Fraser River.

Every fall Mike, Rip and I would repeat our performance in the backcountry after the first couple of sweeps by the cowboy crew had been made. Mike, to my knowledge, never missed a lost cow. He could follow two or three day old tracks up and down the mountainsides, gullies and all the little tributaries leading down off the high alpine country. Like the Mounties, he always got his man or, in this case, *cow*.

Unfortunately, the farthest part of our range did not slope naturally back toward the ranch but instead the terrain led off downhill in all the directions except towards the home ranch. Needless to say this caused us a myriad of problems looking for cattle that went astray. To suggest that Mike merely repaid our investment over those years would be a wild understatement—he repaid it a hundredfold. In those five or six years that Mike, Rip and I did our final tour of the ranges looking for lost or badly strayed cattle, we probably saved the ranch many thousands of dollars. It would be great to report that Mike, Rip, and I faded into the sunset together, but sadly, it didn't happen that way.

Black bears were becoming a real nuisance to our cattle and one in particular, around Red Mountain, was getting away with a lot of young stock. His modus operandi seemed to be to approach up-wind behind cattle which were lying down, preferably lying behind a log in timber at the meadow's edge. If it were a calf or a yearling, the bear could kill them easily. If it happened to be a cow he would grasp one forearm over her shoulder to restrain her while he tore off her udder with his teeth. Then he preferred to let the cow go while he gorged himself on a younger animal. He would then probably lie down on a hillside overlooking the scene of his depredations for a while until the animals would die and the ripe smell would pull him back to another gargantuan feast.

One day Walter Grinder and Frank Peter Meyers were returning from checking cattle on the Poison Mountain Range while riding out of Yodel Camp. They came upon this huge blackie in Red Mountain Meadows who had just killed a young cow and was preparing to devour her. Walter was carrying my short barrelled .30-06 calibre Winchester rifle, but the huge black bear challenged them with such ferocity that the two decided that discretion was the better part of valour. They headed back to Yodel Camp on the gallop. That night when I got in they related their scary story to me. The next morning I was at the kill on Red Mountain Meadows before daylight. When daylight arrived there was no bear to be seen — and no cow either. During the night this big blackie had eaten a whole cow, leaving only head, hooves, and a few hipbones.

To put an end to these depredations, I phoned Frank Richter, the Conservation Department's Predator Control Officer in Kamloops, to come out and get rid of the bear. Frank arrived with many cans of spoiled salmon which he opened and laced with poison. He and I spread these cans around near the various kills. Frank usually nailed the cans to trees to keep them away from other predators. In this case it was all to no avail. After Frank left, I continuously checked the poisoned cans of salmon but they remained untouched. I was at my wits end until I remembered that we had a lot of strychnine in the commissary store at the ranch. This poison was left from the time when Henry Koster had poisoned off all of the wolves in Empire Valley.

I rode back to Empire, picked up the strychnine, and proposed to do my own poisoning. Mike, Rip and I set off one day to bait up all the carcasses that we could find with strychnine. My method was to cut a roast out of a cow or a deer that had been killed but not eaten, fill up the cavity with poison, and then close up the hole again. In this manner we poisoned three or four carcasses. Of course, to Rip and Mike, my actions probably made little sense.

As I rode away from the last kill I suddenly had this sinking feeling that the dogs weren't right behind me as usual. Hollering loudly for them, I rode back to the kill to find that Mike and Rip had torn out the meat plug and were greedily chewing it up. Hoping against hope, I called them away and cleaned out their mouths as best I could. We headed back to Yodel camp, the dogs close behind me now, but had only gone a short way before they began having convulsions. Within the space of half an hour, Rip died in my arms and Mike was nearly comatose. I loaded Mike on the saddle in front of me and raced the six miles to Yodel Camp.

I recalled that Francis Haller, believing he was poisoned at Yodel Camp a few years before, had melted a pound of lard and had drunk that to stave off the effects of the supposed poison—he had in fact suffered a heart attack. Arriving at Yodel, I quickly got a fire going, melted a pound of lard, and poured it down the throat of Mike who was miraculously still alive. Then I started up the old International 4x4. I prayed the decrepit old truck would start, and it did. We rattled and bumped as fast as possible over the twenty-eight miles of four-wheel drive road to home so that I could call a vet. As soon as I got to the ranch, I called Lorne Greenaway and Don Olson in the Kamloops Veterinary Clinic. Don told me to get as many aspirins into the dog as I could and he might make it since he had lasted this long. Mike and I spent an awfully rough night that night as he lay at death's door.

By the next morning he began to rouse himself and before long it was apparent that although still very sick, he would live to track more cattle. The same of course could not be said for my good buddy Rip for whom I shed a lot of tears. She trusted me implicitly and I let her down.

Sometime later we sold Empire, and I took Mike along with us to Kamloops and a home in the city. That first winter of 1967-68, I helped out my friend Sandy McCurrach at Purity Feed, who was fattening several head of cattle on his feedlot at the old Ord Place in Westsyde. Sandy is one of a group of us buddies who had met first at UBC in 1954 and who are all still great friends today.

Mike and I were in charge of determining when the cattle were ready for shipment to the slaughterhouse. Before the cattle could be shipped, however, we would have to run them through the chute and brand or clarify brands and other such things.

Mike would help me put a bunch of cattle into the catch pen leading into a narrow chute which had a Powder River metal squeeze at the chute's end. I would operate the squeeze and catch the cattle that Mike would push down the chute. He was better than a man at this job. He would get behind the cattle in the catch pen and then force the leaders to fill the chute. Then he would crawl under the catch pen fence and jump onto a running board a few feet off the ground which led along the side of the chute. There he could reach through cracks in the boards to nip the last animal in the flanks or buttocks. In this manner we were a great team and could process a carload of cattle in a very short time.

When spring came and all the cattle had been shipped, our job with Sandy ended. Mike quickly became bored with sitting around the house while I looked for a new job in real estate sales. I took Mike out to my father's little ranch on the Kamloops/Vernon Highway at the top of the hill above Monte Creek. Dad would often visit Wayne Everett to help him out at Wayne's feedlot at Monte Creek. Mike would immediately get involved with any cattle work being done there, so much so, that when Dad didn't go down the four or five miles to Everett's Feedlot, Mike would take himself down.

He'd put in a good day's work with Wayne's crew and then get home in time for supper at Dad's place, just like a young cowboy might. People often reported to me that they had passed Mike, morning or night, on his way to his job at Everett's Monte Creek Feedlot. There is a sad ending to this tale, however, as some yahoo decided to run down the dog on the highway. Friends stopped and picked up the sorely hurt dog and took him to Dad's place. I rushed out in answer to a frantic phone call from my mother, Eleanore, to tell me that Mike had been run over. I rushed him to the vet in Kamloops, where Don

Olson examined him and said that we could probably keep him alive, but he would never work cattle again, and might not even be able to walk.

Reluctantly I made the decision to put Mike down, as I knew that he, of all dogs, would not be happy if he could not work cattle. As I recount this story, tears well into my eyes in memory of our unique relationship. Mike was one in a million. They say that a dog is a man's best friend and I recall so many nights camped out with Mike sleeping on one side of me and Rip keeping herself warm on the other side. Those were wonderful days with those two friends anxious to do my every bidding as we worked the cattle together. I can truly say, with all due respect to my wife, that this is true—a dog is definitely a cowboy's best friend.

Chapter Eleven
Bull Shot

We always had trouble getting the bulls back to the home ranch after spending the breeding season in the mountains. It seems as though they had "shot their bolt" so to speak and were drained of energy as well as semen. Every fall we faced the same problem; we would kick twenty-five young bulls out of the Gun Creek Range and into the Tyax Valley one day only to find that a dozen had backtracked the next. That is very frustrating when the weather is closing in, snow is flailing you in the face, grub is running short and the cowherd has already taken themselves halfway home.

Cows are pretty good at getting the message that winter is coming. Bulls, at that point, seem to be brain dead. It is possible that they want to relive the halcyon days of summer and dream of romances consummated with others still waiting. How many of us dream of the good old days when we were young and virile? Sometimes the older of the bulls would just plain refuse to go home. They'd pick a spot where they could hole up in the bush, with water nearby, and figure to camp there for the winter.

Of course, their chances of surviving through the winter in minus forty degree weather, in three to six feet of snow, with ravenous grizzlies coming out of their dens in spring, were somewhere between zero and nil. Don Gillis used to try to burn them out by lighting their winter camp in the jack pine trees on fire, but often that only succeeded in driving them into the next bunch of windfalls and debris.

Another time an old boy went straight out into a small pothole lake and dared me to get near those horns of his with my horse. Horses are peculiar about this kind of situation; they want to keep pretty much away from old bulls on the rampage. It was no use carrying a length of jack pine tree branch with which to beat the bulls because you couldn't get close enough.

Considering that we had to get 80 or 90 bulls home from the summer range you can understand that many of those bulls made our job harder. My solution was to shoot the old beggars through their horns. That got them moving and going down the trail to home. Often, though, the message had to be repeated when they would turn on me and try to gore my horse.

I usually packed a seven shot Ruger .22 calibre pistol in a zippered holster on my belt. That pistol did double duties. Many blue grouse had filled our cowboys' bellies after being shot through the head by old Dead Eye Mack. I was also able to shoot a few bucks with that pistol to complement our rations when grub was short and young bucks were curious and stupid. A few of the old bulls at Empire carried those little .22 holes beside their ears to indicate that they had trouble getting the message. The holes also left another message, not to trust those old boys back in the mountains again and that perhaps the bully beef or bologna abattoir beckoned.

One late fall another old boy holed up on a big open side hill about twenty miles from home and hay, just over the ridge from the Porcupine Creek Valley. I spotted him all by his lonesome while I was flying with Les Kerr in his Super Cub in late

November. In late fall I usually tried to either rent a plane or borrow one from one of our hunters, complete with pilot. Donald and I trucked Bartender out the next day in our Chevy pickup. Bartender was a quarter horse that he had received in trade from Francis Haller for his thoroughbred horse, Whodat. Donald rigged up Bartender, then found and roped the old boy around the horns and dragged him down the hill to where I had found a spot directly below him on this road that led to Roaster Lakes.

The idea was to drag the bull downhill over the road bank and onto the truck parked cross ways on the road with the tailgate open and just square to the bank so the bull could be pulled in. We had even brought a block and tackle to be able to hook the block to the "bull board" of the rack and thus pull the bull into the truck. Bartender was up to the task of pulling the bull against his will, straight down the steep hill to the waiting pickup truck. Once there though, the bull was just too cantankerous and heavy to be pulled into the truck despite Don and Bartender's best efforts with the block and tackle.

We reverted to Plan B, which was to shoot the bull full of tranquilizer while he fought at the end of the rope above the waiting truck. After ten minutes the bull nodded off and lay down, signalling to Don and Bartender that they should yank him into the box. This worked like a damn and I took off for the ranch with the bull safely ensconced inside the horse rack while Donald and Bartender had to hoof the twenty miles to home.

As soon as I got home I backed the pickup to the horse manure pile at the back of the barn where I used a tractor to pull the still sleeping bull out onto the manure pile. I then covered him up with steaming straw and manure, figuring the old boy could possibly freeze to death in his comatose state. That too worked like a damn, and three hours later he was up, gazing truculently around, trying to figure out the startling change of scenery. A few months later the old boy was probably looking at another change of scene as he graced the inside of a bully beef can.

I think I told you about the wicked old bull with the sabre horns, who had laid Frances Koster's horse low in the Dry Farm as she attempted to move him out to a waiting truck. That bull was coerced by Henry Koster and the Grinder brothers into joining a bunch of dry cows that were loaded onto a truck bound for the Ashcroft cattle pens. He was loaded onto the old five-ton Austin cab-over cattle truck, but he could not be coerced off the truck into the railroad holding pens at Ashcroft, so they brought him back and turned him out with the other bulls for wintering at Empire.

We inherited him the next year, along with the ranch. That fall he wouldn't come home. He was probably thinking about the long truck haul to Ashcroft and back, determined not to get a one-way ticket this time. He holed himself up outside the Dry Farm fence for the winter and I found him there in the spring, but he was now pushing up daisies. I salvaged his old skull, complete with horns, and gave it to my father-in-law, Doug Stevenson, who had lights put inside the hide-covered skull and hung it up in a big old fir tree outside the door of his home in Williams Lake. The skull made a great conversation piece. Whenever I saw the head of that old bull he looked as though he was still daring people to "truck with him" through the looks that he cast out of those man-made, malevolent red eyes.

A young bull figured in another pretty good story that ended in success for a large group of our hunters. We had taken three truckloads of hunters and guides down through Shakes Gulch to the Big Churn Creek country. At certain times we would run out of bucks in the first part of the season, so we would give everybody a cook's tour of the ranch, trying to turn up some young bucks.

On this particular trip, we had made a lunch camp at a small lake beside a spring on the Big Churn Creek Flats. In the party was the Jardine family of Kamloops, guests of my Uncle Bob Carson and his son Bobby. We did have several paying hunters with that group, but I can't recall their names. Anyway, we had just finished lunch beside that little lake on the Churn Creek flats, when we could hear a bull bellowing as he made his way in our direction. It seemed as though he knew where he was going and what he wanted as the bellows became stronger the nearer he got to us.

At 200 yards off, I recognized a young bull we had bought the year before from Basil Jackson at Cache Creek, along with several others of his young bulls. Most of Basil's bulls were quite used to being handled by men on foot and this one was no exception. He had been kept in a pen below our horse barn all winter and he was a lot tamer than any of the other bulls. I had named him Shorty. Our hunters were all getting a little apprehensive as the bull neared our lunch campsite.

"Don't worry," I said, "I'll take him on."

I went out to meet Shorty, bellowing as I went. When we were separated by ten yards I began pawing the ground as I lowered myself onto all fours. Shorty quickly responded by bellowing and pawing the ground as well. This must have made quite a spectacle for our wide-eyed hunters. Finally Shorty and I locked horns, at least I grabbed him by the horns, and we wrestled a bit as we were used to doing back at the barn. Then I jumped on

Shorty's back, facing absurdly backwards, and used his tail as a steering wheel.

"Follow me," I said to our assembled hunters. Off we went, across the flats, towards Churn Creek. Our hunters all piled into their trucks and followed me and Shorty across the flats about a mile to where I jumped off and peered over the edge and down into Churn Creek. There, sunning himself at the bottom of the valley, on a small ridge about six hundred yards below us, was a big four point buck. Our hunters lined themselves up on the edge of the bluff and alternately proceeded to pepper (or salt) the hills around the buck with lead. However, the range was such that at 600 plus yards, nobody could bring him down. The buck, for his part, couldn't figure out where all the noise was coming from. All he knew was that clods of dirt kept appearing in front, behind, above, and below him.

Finally, one of our paying hunters, sporting an ancient .303 caliber British—an old army rifle—knocked the buck dead, when he lifted his open sites to account for the extreme range. The old .303 had probably seen duty in the hands of a Canadian sharp shooter in the First or Second World Wars while with Canadian Army troops. That rifle was ideal for shooting long range. This certainly wasn't good advertising for the scoped Winchester and Remington .270s, .308s, and .30-06 caliber rifles which just couldn't come to grips with that long distance shooting.

It remained then for me to pull a quarter mile plus of three quarter inch rope down the hill, hook onto the buck's horns and drag him up the hill with the four wheel drive truck. At one point the rope was tightened a hundred yards above the valley in a straight line to the buck. He was a beautiful trophy buck, brought to bay by a clowning cowboy on a broke to ride bull.

Chapter Twelve
Horses I Have Known and Respected

Our horse herd at Empire Valley numbered about forty head of saddle horses and work mares when we purchased the ranch in 1956. When we sold the ranch in 1967 the herd now numbered about 150 head of working cow horses and many colts plus a few draft horse mares. A friend, Billy Wraighton, gave us the use of a thoroughbred stud named Dansky, who had bowed his legs as a two-year-old racing on the Hastings Race Track in Vancouver and needed R and R on an interior ranch.

With that beginning we eventually ended up with a large remuda of good working cow horses and packhorses. The packhorses were the result of breeding Dansky and another later Wraighton stud named Mr. Worth to our daft mares. . .

We also purchased unbroken saddle horses whenever we could get a good deal. In such a deal we bought Duke (my old reliable cow horse) from the Alkali Lake Ranch that was owned by the Riederman family. We purchased Nifty, Rusty, Blue, Sheba and a few others from the Circle S Cattle Company at Dog Creek where Tommy Desmond was the cow boss. He had professed to me that he had an excess of unbroken saddle horses.

The saddle horses I worked with on the ranch were Fever and Dolly—both came unbroken with the ranch purchase—as well as those we purchased— Duke, Rusty, Buckles, Salty, Babe, and Shifty—my best horse. Sometimes I rode my dad's mare, Dude—the one I was riding when I got a concussion just before I was to marry Liz.

Fever—with Coast Range behind

Don added Blue to his string, his daughter Monica rode Sheba and I kept Nifty and Rusty. Tommy Desmond told Sophie, one of the Riederman sisters, that I was looking for a good cattle horse. I went to Alkali Lake, and when we drove up onto the range, Sophie called all the loose horses. They came running to her call and to her bucket of oats. She put a piece of bailing twine around Duke's neck, the one of the herd that I liked best, and she led him down to the ranch with nothing more than the bailing twine. So I knew that he was gentle and used to coming to her summons when she'd arrive with oats.

Nifty died from getting loose and eating too many oats at the Dry Lakes Cow camp during the night; she was a great horse too. I got Shifty from my brother-in-law Don Gillis, along with Molly. He told me by phone, when he was still at the Glen Walker Ranch on the Coldwater River in Merritt, that he had a few good ones for me, so I went down and got two from him. They were unbroken at the time. I brought Babe, a big sorrel quarter horse mare that Art Hay, the manager of Nicola Stock Farms, had given me. Babe became my wife's horse at Empire but was my cow horse in Merritt. I had a friend, Alfie McKnight, with me the first time I rode Babe out of the corral and she bucked me off into a cactus patch. Alfie was quite concerned but I told him not to worry about me, *just catch that damn mare*.

Devil was a gelding from the Diamond S ranch that was given to me by Colonel Victor Spencer of Spencer Stores in Vancouver when I was a kid, because he was fond of me and my dad. He also gave me the complete set of Tom Sawyer books for Christmas. I won the best horse and rider on Devil at the Merritt fall Fair.

Devil was a gelding by Emperor Hirohito's white stallion. It was reputed that Colonel Spencer's ranching partner, a man named Frank Mackenzie Ross—later the Lieutenant Governor of BC – had brought the stallion back from Japan when the war ended. Supposedly the white stallion was obtained from Hirohito in Japan.

One of my jobs when my Dad was managing the Diamond S was to water the big white Arab stallion. His jaw bone was about two feet higher than my head when I was about 12. I also had to clean his stall. I found him scary at the time. Devil was a beautiful black, half Arab, with a white stripe on his face and four white stockings.

The Colonel once took me, my brother Dunc and a friend, Eddie Lehman, to the Calgary Stampede and bought us each a brand new bridle. We stayed at the Palliser Hotel, in a room with rugs four inches deep, that we wrestled on. At the time this was a very special treat for us country boys. Spencer owned three ranches: Earl's Court at Lytton, the Circle S at Dog Creek, and the Diamond S at Pavilion. Later the Colonel's daughter, Barbara Spencer, owned the Circle S with her dad and they employed Frank Armes as the ranch manager. His son Gordon lived there and, as coincidence would have it, later married my wife's sister, Rhona. The Colonel's son, Victor Spencer Jr., bought the Douglas Lake Cattle Company near Merritt and later they sold it to Chunky Woodward of Woodward Stores in Vancouver.

Because we had so many horses, naming them was a bit of a problem, so Joe Matson, one of our cowboys, began rhyming names to match one of the standbys. As an example, we had Dolly, a mare we inherited with about 20 others when we bought Empire Valley in 1956. Dolly, therefore, was one of the oldest horses on the ranch. After I broke her, she was best known for her ability to "pick them up and put them down." It was a quality that came in handy whenever I was getting from point A to point B in a hurry. The names Holly and Solly, Joe added to the Dolly line. We had other lines named on the rhyming scheme as well, such as Handy, Dandy and Brandy.

This story is about Shifty, who arrived in 1957. When I brought Shifty and Molly home from the Coldwater they had had just gone through a pretty rough winter. After a few months of hay and oats that we grew on the ranch, they became fit enough to break. The following is a true story and could only happen to one horse in a thousand.

The day appointed to start breaking Shifty was one of those beautiful sunny spring days when I loved to ride out to the riverfront to check on cows heavy in calf. On those lethargic days it was nice to pack a lunch and find a big rock to put my back up against while I soaked up the spring sunshine and looked at the miles and miles of undulating rangeland surrounding me. Needless to say I often had a cat nap on those first sunny days of spring as I watched the cows grazing hither and yon and the deer sliding gracefully in and out of the park-like gullies.

At any rate, it was one of those days when I led Shifty out of the barn and into the breaking corral where I tied her to a post while I "sacked her out" (lightly slapping her with a sack to get her used to sudden surprises.) Within an hour I had her saddled up without an attempt by her to throw off the saddle. Soon I put a bridle over her head and a snaffle bit into her mouth which caused her to jerk her head back and eye me suspiciously. Then I attached the bridle lines to saddle stirrups and with a long line I was able to drive her like a draft horse. Next I tightened up one of the bridle lines which would cause her to turn in circles. I then alternated tying one bridle line to a stirrup so that this caused her to turn in circles in that direction because of the pressure of the bit in her mouth. I then left her to her own devices while I went for lunch at the cookhouse.

After lunch I stepped into the stirrups again and again until I gradually eased myself into the saddle. I practiced getting on and off the saddle without triggering the normal bucking horse response that many horses resort to when they feel the weight of a rider on their backs for the first time. After about an hour of this, and turning her head with the lines, I could ride Shifty around the corral in circles. Then I changed the pressure on her mouth to correspond to the pressure on her neck. The left line pressed against the neck meant turn right and vice versa with the right line. Within an hour we were basically in the first stages of neck lining around the corral.

At the same time I unhitched my lariat and began swinging it and throwing it, while we plodded around the corral. Shifty seemed so amenable and gentle that I got up enough courage to open the gate with the expectation that I could ride her out and show anybody who happened to be around what a fine cow horse she was going to make. I felt emboldened enough to then ride her out into the main ranch yard below the buildings. Shifty seemed to be very intelligent and quite responsive.

It was about then that Cuff, who had been watching the whole process for these last few hours, decided it was time to intervene. Things were obviously going too well and it did not appear that he was going to see some violent action with me ending up in the corral dust. So he quickly slid in behind her and took advantage of this naive young thing by nipping at her heels. This did cause a bit of a ruckus but not nearly what he had intended.

Before he could really savour the rewards of his attempts to unseat me, I had unlimbered my lariat and, without really thinking about what I was attempting, I kicked the mare in the ribs like a broke horse and started after the dastardly dog. Perhaps the nip on the heels had alerted Shifty to the opportunity for revenge because before any of us knew what was happening,

Cuff was headed flat out for the cookhouse to get under the porch and Shifty, with me swinging the rope over my head was in hot pursuit. Perhaps some atavistic impulse was guiding her more than I was for we hit the corner of the cookhouse going flat out. Cuff was barely able to round the corner of the cookhouse before we arrived, hell-bent for leather. At this point things were a bit out of my control, and the biggest problem presented to me was not trying to stay seated on a green, barely broke, mountain pony going flat out. It was rather how to duck under the metal dinner gong while Shifty performed a kind of right angle turn around the corner of the cook house that would have made a barrel racer proud. Shifty made that turn without losing a step and Cuff made it under the cookhouse porch before I could get my rope on him.

For myself, I managed to narrowly miss clanging that iron crowbar dinner gong with my head. We must have made quite a sight, but only our startled cook, Clara Paulson, was witness to this amazing feat by a mare saddled for the first time only hours before. In retrospect, it was getting close to dinner when this all played out; had I hit the gong with my head at the speed we were going, we would likely have raised the whole crew out of the bunkhouse to watch the boss get some sense knocked into him. With that auspicious start Shifty became my main cow horse. She had a colt by Dansky called Swifty who was nothing like his mother. Dansky's colts were all very tough to break and some just couldn't be broken to ride.

When we sold Empire in 1967, we sold a black mare called Black Page and a big bay gelding called Mr. Empire that had been sired by Dansky, to the Perry and Hook Rodeo string out of Kamloops. Rodeo riders loved Black Page because she wasn't that hard to ride but could always be counted on to score 65 or better for the judges. (A score of 65 is a better than average score. Scores of up to fifty points are given to the way a horse bucks and up to 50 points are granted on the skill of the rider.)

Mr. Empire, on the other hand, was ridden only a few times. As far as I know, he was the only horse ever to put the great Canadian Champion, Kenny McLean of Okanagan Falls, on the ground (twice). When Dansky's colts were ready to break at the age of four years I had hired a horse breaker because there were too many colts for Don or me to start. This horse breaker named Norman Alphonse, from the Chilcotin country at Alexis Creek, was the brother of our first cowboy cook named Clara (Alphonse) Grinder. Norman got some of Dansky's colts going but gave up on Mr. Empire because he was just too mean. Next I hired Don McDonald and Chris "Cactus" Kind to break the remaining colts but again they gave up on Mr. Empire as well because he almost killed Don McDonald in the breaking corral. He had charged Don and only a gigantic leap up to the top of the fence saved Don from the flying rear hooves that could have done him in.

When we left Empire we left behind quite a few mature horses and colts. Bob Maytag must also have discovered that one of those that we left was obviously not easy to break either, as the horse ended up at the National Finals Rodeo in Denver Colorado where he was crowned the top bucking horse in North America. He was called Two Bits, in reference to the 25 brand which all our livestock carried. The number 25 was branded on the left side of our horse's shoulders and on the right hips of our cattle.

For my part, I was not going to give up on Shifty's only colt, a son of Dansky that I had named Swifty. Because he had Shifty's blood line, I was sure that he would make a great cow horse given time and patience. Things started out fairly well with Swifty, partly because I had Francis Haller take a short lead from Swifty's halter to the saddle horn of his own horse, Spooks. Since we were riding the riverfront in springtime, checking cows and calves that had been turned out of the hayfields to graze on the early spring grasses of the Fraser Riverfront, we could start from the ranch headquarters with Francis keeping Swifty under fairly close control. After an hour and a few miles of road we would cut out onto the rangeland for our daily inspection tour. Once we were onto the rangeland proper, Francis would return me my halter shank and I was on my own with Shifty's gelded son, who would someday make me proud, I was sure.

However, for four or five days in a row, Swifty would go into a bucking frenzy over some provocation, like a cactus jumping up onto his belly, or a blue grouse breaking out from a sagebrush under his nose. On this particular day we were just outside the Bishop Ranch hayfields and veering off the main road and onto the rangeland. A particularly nasty cactus sprang at Swifty's underbelly and away we went. We were a pretty good match till Swifty bucked himself sideways and backwards over a ten foot embankment back onto the main road. We ended up in a heap onto the relatively soft clay roadside. While Swifty struggled to regain his feet, I clambered on board again for the rerun up the bank and full tilt across the bunchgrass side-hill. Francis reined in beside me and we got Swifty simmered down just before we happened onto one of the numerous pothole lakes abounding on this river range in springtime. The pothole seemed like an opportunity whose time had come so I spurred Swifty into the lake and round and round the acre sized pothole until he was in a sweating lather.

Good, I thought, *this will knock some of that piss and vinegar out of you, and I won't have to spend every step waiting for the next bout in our daily bucking matchup.* He was pretty tired, there was no doubt. I was particularly pleased as I pulled up alongside Francis who had been enjoying this particular episode, as only one who has seen and done it all himself could really enjoy.

Francis, Mack and Don (courtesy Bob and Allan Donnely)

"That'll teach the SOB," I huffed to Francis because I was a bit winded myself. However, before Francis could come up with a suitable comment the Son of a Bitch lit into it again—perhaps my references to his esteemed mother set him off. Whatever the reason, that Son of a Bitch that I had thought was completely winded and had been taught a good lesson, managed to dump me face first into a cactus patch. Luckily Francis's horse was fresh and able to quickly catch the winded Swifty who had set off back to the barn with halter shank and bridle lines flying. We continued on that day without any more shenanigans. Apparently the rest of day was seen as a truce by Swifty—but only that day.

I think every time I rode that horse, he waited for that one golden opportunity to put me on the ground. Fortunately for me, once was enough, and I was always ready for him after that spring swim that we had enjoyed together. It is odd to think that a top cow-horse like Shifty should have borne a son like Swifty, but not so odd, I guess, when you compare this mare and gelding to some human counterparts.

All this goes to show that (a) I wasn't as good a bucking horse rider as I thought I was (b) don't ever count out one of Dansky's offspring when it comes to dumping good cowboys, and (c) Swifty proved that none of Dansky's colts ever did make a good cow horse, even when his mother was the best cow horse I ever rode.

Chapter Thirteen
Falls I Have Taken

All cowboys, if they stay in the business long enough, are going to get piled. The more hours he spends on a horse's back, plus the roughness of the country and the unpredictability of many different cayuses, all contribute to the "whens" not the "ifs" of getting bucked off. However, no cowboy worth his salt ever wants to be referred to as having "fallen off" his horse. In cowboy vernacular, you can get piled, bucked off, throwed, stomped, crushed, scraped off, or catapulted, but to be guilty of "falling off your horse" would be tantamount to handing in your chips at a poker game because you just were not a real cowboy.

Consequently, I've never fallen off any ranch horse, but I have been piled off quite a few—as I say, the longer you stay cowboying, the more your chances increase. The object, of course, is not to get hurt too badly when you get thrown, and here is where the skill part comes in. Just as a TV wrestler must learn to take a pratfall and make it look real, so must a cowboy learn to take a fall, but make it look normal when he rebounds to his feet and back into the saddle without injury.

Any of you who have watched rodeo cowboys get bucked off in a staged arena event have witnessed professional rodeo cowboys demonstrate amazing athletic ability in turning in what could be a disaster into, "Give him a big hand, folks, that's all he will receive," from the rodeo announcer.

The ranch cowboy does not receive the adulation of the crowd when he gets piled while going about his daily routine and he never has the option of falling on six inches of carefully manicured topsoil to cushion his fall. Rather, when he pulls his aching bones back onto his bronc, he congratulates himself that one more time he has beaten the odds and survived the rocks, trees, cliffs, and cactus that come up awful fast in that short descent from the kingly position astride his mount. He counts himself lucky if he has bruised only his ego, considering the sometimes compromising blow to the body's bones, especially if he hits the deck at speeds of up to 30 miles an hour. Of course, if he has a chance, a cowboy will hope to land on his shoulder and do a quick roll to prevent injury to his body from the sudden jolting stop.

STUD HORSES AND THE RESULTS

As I mentioned, a friend, Buddy Wraighton who trained race horses for the Hasting's Park Race Course, had loaned us a stud named Dansky who had bowed his tendons as a three year old on the track. Dansky was the first of two thoroughbred studs he had loaned us. When he took Dansky back to the track the thoroughbred stud became one of Hastings Park's all-time money producing studs while standing at stud on the track. Buddy then offered us the replacement of another thoroughbred stud who had also bowed his tendons. Mr. Worth, the second stud he loaned us, was a British bred stud by Mehmander, and he stood six feet at the shoulder. Mr. Worth was a gentle amiable bay whereas Dansky was an athletic, mid-sized, red coloured sorrel with a black mean streak.

Don Gillis had to ride Dansky home bareback with only a halter shank to guide him when he trucked the big red stud to the ranch for his first date with our mares. Don had gotten his truck stuck in the spring runoff, which often caused our road to turn into an impassable hell. Don didn't like walking in his cowboy boots so he chose to ride Dansky bareback. We never did try to break Dansky to be a cow horse. Of course, he had already been ready for the track before he bowed a tendon as a three year old. This was why Buddy Wraighton chose to send him to Empire for a complete rest with bedroom privileges. Dansky was very mean and I am still in awe that Don rode him home bareback with only a short rope around his neck. As I mentioned earlier, Dansky sired quite a few top rodeo bucking horses.

Mr. Worth, on the other hand, was the complete opposite of Dansky. He fathered a passel of real gentle cow horses that only occasionally caused me to bite the dust. Because of his disposition, one spring day when the sun beat down on the river front and caused my riding juices to start flowing, I decided to make Mr. Worth into a cow horse. A few days of training in the corral and he was ready for bigger things. He didn't really feel as though he would make a cow horse and he didn't seem to want to get his long legs into a ground-eating walk but he was gentle and the chances of him dumping me were slim. So it was that I headed up to the Gap hayfields to check on calving cows.

Having found all to be in order, I headed him back to the barn for lunch. On the way, as Mr. Worth dawdled along, I got to thinking; *this horse has to have more gears than just bull low*. So I punched him in the ribs a few times and was rewarded with an awkward, ambling trot. Emboldened by the speed of this second gear, I punched him again in the ribs to see if he indeed had a higher gear. Mr. Worth responded again with a shambling attempt at slow gallop.

"Damn it," I said to him, "how the hell are you ever going to make a racehorse if this is the best you can do?"

Again I really drove him in the ribs whilst he maintained his slow gallop. This time I was caused to understand what gives jockeys their thrills on the track, for the big fellow could really stretch out.

Whee, this is fun, I thought shortly before Mr. Worth and I bit the dust at thirty-five miles an hour. We had been going flat out across a hayfield, which still had patches of snow, ice, and frozen ground. Hidden under all of this were indentations or "marks" as our irrigators called them. These marks caused the irrigation water to flow from the ditches through the dry areas of the hayfields. Sometimes it seemed as though they could make the water flow uphill. Here, in early spring, the freezing and thawing had made these marks slippery on the top quarter inch and frozen solid underneath.

Mr. Worth and I hit one of those slippery spots and when the big awkward footed fella fell flat on his side, he dragged me along with him for about six yards, bouncing over—I swear—a dozen of those irrigation ridges. My right leg was pinned under the stud horse for the brief wild slide. It is difficult to explain to you what damage a twelve hundred and seventy pound stallion, using your leg as a shock absorber, can do to a skinny cowboy leg.

I gave up on riding the big fellow that day as Mr. Worth and I slowly limped the rest of the mile and a half to the ranch. I was laid up for two weeks. This was the only time that I had ever been injured on the home ranch at Empire. That fall shows that the stud had more brains than this cowboy. The only good thing about it was that I had my beautiful bride, Liz, to wait on me and my two year old daughter, Lisa, to commiserate with me.

Liz was a great consolation to me as she administered to my aches and pains and Lisa seemed to get inside my skin and feel how I felt.

As an example, one day that summer, I was working around the ranch and came home at lunch to be told by Liz that the Servel fridge on the back porch had stopped working. These fridges were fired by kerosene and were very temperamental. The whole idea of burning kerosene to create cold was, first of all, too much for this cowpoke, but I did understand the elementary things like filling the tank, trimming the wick, and cleaning the burner.

To complicate things, on the 100 degree Fahrenheit hot summer day was the fact that our deck, where the fridge resided, was absorbing all of the sun's heat. Inside the porch the temperature radiated up to what seemed like 120 degrees Fahrenheit. Fighting with the Servel while lying on my side on the rough planks of the porch deck was not very conducive to a healthy temperament.

Normally in these situations, I can fire off a few healthy curses to alleviate my frustrations. On this particular day I was doubly stressed as the Servel was not cooperating. As Lisa was sitting beside me on the porch floor, I couldn't let loose with a few expletives to lessen my stress. Lisa, however, could sense all my pent up emotion, it seemed, and really broke me up during a particularly frustrating moment when I was attempting to get the Servel fired up again.

"It's a summumabitch, isn't it Daddy?" she offered astutely. Lisa had literally broken the ice for both me and the Servel, which now decided to get back to doing exactly the job for which it had been designed. (We did have electricity on the ranch but the light plant, which was on all morning, was turned off at noon, back on at 5:00 pm and off again after nine o'clock at night. This would have made it difficult to keep perishables at a constant temperature, not to mention keeping the frozen food from defrosting.)

Mr. Worth begat a lot of big, beautiful colts and fillies but one in particular sticks in my mind. I dubbed him Brandy, part of the Brandy, Andy, Handy, and Sandy line. He was a beautiful big bay gelding with a split ear that he had picked up somewhere while at his mother's side. He was one of a dozen colts that I had hired Chris Kind[3] and Don McDonald to green break—getting them used to the saddle and bridle and seating a rider in a corral—as I was too busy finishing the horses I had started. I had Brandy pretty well started on his journey to be a cow horse, in a string of a half dozen colts that I was riding in the fall of 1963.

So it was that Brandy was ready to go along with my string of saddle horses on one of those eventful days that seemed to keep cropping up in my life as a cowboy. I had Walter and Henry Grinder and Norman Ned with me as we searched the swamp meadows, gullies, creeks and alpine ranges around Red Mountain and the Red Mountain Meadows during one of our last fall gathers. On this particular day I had sent Walter Grinder down to Starvation Canyon and had Henry and Norman riding the Red Mountain swamp meadows up above.

For myself, on that particular day, I had chosen to ride Blue, out of Don's string of horses, from our Yodel Camp up onto the Poison Mountain, Red Mountain area. I always liked the view up there but the cold winds were blowing as I attempted to ride the toughest area myself. I found a few hungry lost cows hiding on a scrub-covered hillside on the south side of Poison Mountain and thought I'd drive them along the Poison Mountain Road and from there on to Red Mountain itself where there was no road and very few trails. I was hoping to find the one obscure trail down off the mountain to Roaster Lakes near Yodel Camp.

All went well until I hit Red Mountain and a savage snow squall blew in. At 7,000 feet that wasn't too unusual for this time of year, but it was doubly unsettling because it was also rapidly closing in on darkness at approximately 4:30 PM. When I couldn't see the cows in front of me anymore, I began to think more and more about just getting off that damn mountain. Finding the entrance to a very dim, unmarked trail down off the mountain would have been difficult at the best of times, but blinding snow and darkness made it virtually impossible. I abandoned my cows and tried to get my trail instincts attuned to this unhealthy situation.

After groping around aimlessly for half an hour, I realized that I wasn't going to hit the trailhead and now had no idea where I was. I decided to head Blue downhill to look for shelter in the hope that we could somehow blunder through to Roaster Lakes. That was not to be, however, even though this long legged blue mare was adept at jumping over windfalls and wanted to get to Yodel Camp and hay as badly as I wanted the shelter of a warm cabin.

In full darkness and with pan sized snowflakes falling, we pulled up short in a spruce tree meadow somewhere down off of Red Mountain. Since we were hemmed in, I wasn't worried about the mare and so I unsaddled her and turned her loose to forage in the lush meadow grasses, covered with six inches of fresh snow. Dead twigs, still attached to the base of a large spruce tree, combined with some dry grass and dead needles that I was able to find under the tree gave me a start for a pretty good fire.

So I guess we're here for the night, I thought. *We'll just have to make the best of it.*

Fortunately, I was wearing my eiderdown jacket and chaps on my legs so I felt that I could weather through alright. The temperature was barely freezing and something my dad had mentioned to me one time when he was in a similar situation with some hunting friends over on Milk Creek near Clinton, came to mind.

We built two fires and three of us lay down between them. We had a piece of canvass covering us and about every hour someone would holler, 'Roll' and the guy on the outside had to get up and rebuild the fires.

With this in mind, I decided to build two fires, for there was ample dead wood lying around in the form of windfalls and dry branches at the base of the numerous spruce trees. Soon I was quite comfortable and fell asleep only to awaken in order to rebuild the fire and observe Blue's eyes peering at me out of the darkness to make sure I didn't somehow leave without her.

At times I was aware of extra smoke and heat but that only caused me to enjoy my bivouac more as I reclined in the arms of Morpheus. Along about daylight, Blue began whinnying, and I figured that she thought we had best be on our way, so I saddled her up and rode uphill looking for a viewpoint to figure out where we were.

After climbing for about 1,000 feet we got ourselves on a ridge where I could see some landmarks. *Hell,* I thought—and I wasn't referring to our bivouac—we were near Pete Caldwell's hunting camp at the headwaters of Lone Cabin Creek. We readjusted our mental compass and headed in the opposite direction, towards Roaster Lakes. There we could find a four-wheel drive road that would lead us to Yodel Camp, situated at the base of Red Mountain. Before we left, I took one last look at where we had been and was tickled pink to find out that our little campfire was now about five acres in size as the fire had burned under the snow in the dry grass and had reached up to any spruce tree in its path.

Good, I thought, *more good grass for the cattle next summer. I hope it cleans up all the windfalls in the whole darned country.*

Now Blue and I really had our work cut out for us going downhill as we were faced with an impenetrable windfall covered side hill that we had somehow manoeuvred ourselves onto. I dismounted and led Blue down off the mountain. I scrambled over the three and four foot high windfalls obstructing our path. Blue, when she was given a little slack on her lead rope, would launch her front legs into the air, jump over the windfall, and slide to a halt in front of the next bulwark of windfalls. In this manner we negotiated the steep side hill and found ourselves close to the Roaster Lakes Road.

"Nice work, Blue," I congratulated her. "We'll back to Yodel Camp in time for breakfast."

However, when we got to Yodel Camp, Walter and the boys had just finished breakfast and were headed out to look for more cows. "Hey," I whined, "wasn't anybody going to look for me?"

"You're the boss," Walter offered. "We knew you could look after yourself. Besides it was too dark and too snowy to go out looking for you."

"Thanks a lot," I whimpered as I prepared my own cold breakfast. "Catch me up Brandy, please, and I'll be on your trail as soon as I can convince my belly that my throat's not cut."

Now Brandy had only recently started on his route to be a cow horse and the thought of the others having left him behind caused no end of trouble as I saddled him up and turned Old Blue loose for her just rewards. In the space of a few moments I jumped aboard a prancing, dancing and pawing Brandy as we headed after the boys. Brandy was grained up, fresh and raring to go and so it wasn't long before we were into high gear with me more or less hanging on for dear life as we rapidly shortened the distance between Brandy and his departed buddies. The trail, which was really the remains of an old narrow road built by Henry Koster with a little D-2 Cat several years before, took a sharp bend around a rocky outcropping along the creek as it meandered down to Starvation Canyon. At that point the recently tightened cinch must have bitten into Brandy's ribs a bit too much for his liking for he launched into a series of mind numbing crow hops while he negotiated the steep turn at a scary speed. The upshot of this was that he caught me with my guard down and when he zigged, I zagged.

In my defence, I was weakened by my solitary sojourn on that snow bitten side-hill meadow half way up Red Mountain meadow. So I fell off—or rather was *thrown off*—into a pile of boulders and debris pushed aside by Henry's D-2 Cat. It all happened so fast that I really thumped into the rocks. *Hell*, I thought, *this time I really bought it and all this pain indicates that I am mortally wounded.*

However, when I got my wind back, I discovered that I was not only alive, but I could move, so I wasn't "all broke up." Within minutes Walter arrived back up the trail leading a chastened Brandy.

"Did you lose something?" he said.

"*Damn, these guys really don't empathize,*" I thought. I reminded myself that, back on the ranch, Lisa would understand. She always did.

—

*3 Chris Kind is the author of five books, the one on Empire Valley and the Gang Ranch is a best seller.

Chapter Fourteen
Big Bucks

Empire Valley became the legendary hunting ground for big mule deer bucks thanks to the efforts of two sportscasters named Ted Peck, of the television program BC Outdoors, and Mike Crammond, who was a Vancouver Province newspaper sportswriter.

Once Mike's stories appeared in the Province newspaper and Ted's TV shots of huge antlered bucks hanging on our hunters' hanging rack hit the sporting headlines there was no stopping myriads of hunters from heading to the Cariboo-Chilcotin hunting nirvana.

It was our own fault, of course. We didn't have to accept the requests from the media to film the Empire Valley herds. In our own defense, we were a long way from civilization at the terminus of a long and winding road. It seemed like a good idea at the time to liven up our daily lives and show off this beautiful country a little. We didn't anticipate the results of "showing off" too well.

My job as cow boss took up most of my time, but in the fall I would sometimes find the time to do a little game guiding. It was a lot more exciting and I was a licensed guide. A few events concerning the media stand out in my mind. On one occasion, the guides had done their best to drive a big buck down the front sites of Ted Peck's rifle for his TV Outdoor Show, but he couldn't, as the saying goes, "hit the broad side of a barn at 50 paces."

After numerous attempts his film crew advised us that they could edit the film as though Ted was actually shooting one of the humungous bucks that we ran at him. In fact the dead ones were killed by our more experienced paying hunters. All legal, of course, as Ted's film crew explained. When they went hunting for pheasant they would throw a dead pheasant in the air and then run the film backwards, as though the bird were being bagged by the great TV Hunter.

Ted was a great TV personality with plenty of good TV shots but, from my perspective, not a great hunter. Mike Crammond of the Vancouver Province Newspaper was not intent on killing anything for the sake of the cameras. He was more interested in reporting on the herds of mule deer streaming into our ranch hayfields at first light of day. Of great benefit to us was that Mike realistically portrayed the excellent protection that the herd received from our family held ranch.

There had, of course, always been mule deer in Empire Valley. The ranch headquarters were situated at the tip of a huge, natural funnel stretching one hundred miles back into the Coast Range. The ranch headquarters itself was the only part of the funnel to contain the necessary ingredients for the deer's winter range. It had a relatively moderate snowfall and cold weather compared to the mountainous back country situated above the Relay, Tyax, Gun, Noaxe, and Churn Creeks. These were but a few of the mule deer summer ranges. Empire Valley also possessed semi-desert vegetation which made for good mule deer winter range habitat. There were also steep Southern exposed slopes above the Fraser River, where the wild winds of winter kept the hillsides bare, not to mention six hundred acres of fodder producing hayfields that the deer grazed on till hard freeze-up.

When winter began in earnest they crawled through our stackyard fences to fill up on second crop baled hay. This was a bit of a nuisance because they ruined more hay than they ate, but we put up with these antics in the interests of keeping the herd healthy. All these necessary ingredients were in place for wintering mule deer and so, when Henry Koster and the BC Game Department poisoned off the wolves which had kept the herd thinned down, then the numbers of mule deer really exploded.

Our game guiding was kind of forced on us because the expanding population of deer was eating us out of house and home. Also, in the early days of our stewardship of this tremendous cattle ranch, we had an expanding population of deer, cattle, and horses. Our families were growing, and we had a nucleus of experienced, dedicated cowboys and farmhands but we did not have one essential ingredient—hard, cold cash.

We had small operating loans with the Bank of Montreal in Clinton to almost tide us over from one set of cattle sales in the fall to the next fall sales. The loan was intended to pay for the operating costs of the ranch, including wages, gas, machinery, food and other expenses. Unfortunately, there never seemed to be more than just enough money to pay off the year's operating loan from the bank, as well as Koster's mortgage. Our cattle sales barely covered these large debts. This left nothing for us partners to spend on holidays and recreation. Taking in cash-paying hunters bridged that gap and as well, kept the deer herds under control.

We averaged about one hundred reputable hunters every fall with approximately twenty of those being American hunters. Out of country hunters cannot hunt without a licensed guide, of which we had several including Donald, Dunc, Art Grinder, and me. All told, we harvested about two hundred mule deer bucks per year. I still count as friends many hunters whose primary interest was in getting out into the wilderness to see some of British Columbia's best scenery and wildlife. For them, the venison was a bonus. If one throws in all the grass fed beef for dinner, the camaraderie and the good home cooking, then our hunters were presented with a recipe that kept them coming back year after year until that time when we eventually sold the ranch. For the first half dozen years of guided hunting we charged $90 for a three day hunt which included room, board, transportation and care of the venison. In later years the price climbed to $150 for a three day hunt, a bargain even in those days.

There must be many Empire Valley bucks' trophy heads hanging up in the recreation rooms of BC and Washington State hunters. All of our hunters, to a man/woman, went away satisfied. They didn't always want to kill their limit of two bucks, but they definitely took home all the venison that they wanted, and had a great outdoors experience to boot.

We didn't hunt does or fawns; we only hunted the bucks that streamed in behind the does from October 15 onward to the end of hunting season in late November. We generally ran about three or four guides with each guide handling approximately two hunters. Some liked to hunt on horseback and others preferred working from four wheel drive vehicles, which took the hunter into the right spot for a sure footed guide to push the bucks out into the open for clear shooting. We guides eschewed the red coloured clothing of a hunter for the more earthy brown tones of the bunchgrass covered hills we operated in because we felt the bucks could more easily spot us if we had bright clothing. That led to a few close calls with hunters peering through their four-powered scopes long enough to satisfy themselves that the moving object was not a deer. None of our hunters ever fired a shot at one of our guides.

That didn't include poachers, unfortunately. The way we set up our guided hunting operations protected does, fawns and guides from trigger-happy poachers. Even splitting half of our privately owned ranch lands for the purposes of guiding our hunters and the other half for the use of public hunters, didn't solve a deer availability problem for the public hunters. Because of people like Bill Federation and his BC Safaris, we had the makings of a range war developing. Bill would bring busloads of meat hunters up from the Coast every weekend. He would camp them alongside the Dry Farm fence which separated our privately held lands from public lands. He would then point out the deer on our side of the Dry Farm fence and away they would go "enfilade," spaced one hundred yards apart into our privately held areas.

These meat hunters usually carried $50 army rifles and had paid Bill $35 for the bus trip to Empire Valley. Everything they required for hunting, including food, was packed in their duffle bags. Because the exercise for them was to bring back some meat for their families, they would shoot anything resembling a deer. Fawns by themselves looked pretty much like a mature deer at a hundred yards so these amiable Bambis were slaughtered out of hand. The does were slaughtered next as they were used to us operating around them and so they had no real fear of man. Anything that remained of the herd frequenting the public hunting grounds that was not killed was driven to seek shelter further into our private lands and the area we used for our guided hunters. Soon the meat hunters were scrambling over our 'No Trespassing' signs to chase down anything that moved.

Not all the actions came from Safari's hunters, of course, and one incident reflects on the dangerous situation that was created when the only surviving deer were quartered on our protected areas. Since the Dry Farm Pasture's fence stood between good hunting and bad hunting, between our hunting and public hunting, I had posted Joe Matson, one of my cowboys, to patrol the area on horseback to keep out any undesirables. Joe took his job seriously and when he encountered seven Burnaby steelworkers encroaching where they were not wanted, he advised them to pull stakes. This they were not prepared to do. They were adamant that, "One ranch could not possibly own all this land and they would damn well hunt if they felt like it!"

(We had heard that one over and over again.)

Joe took exception to that, and when asked how he could enforce his dictum being unarmed, he took down his lariat and proceeded to swing it over his head. Knowing Joe, he probably advised these steelworkers that he would "drag them out." This action caused three of them to charge Joe, who was mounted on my wife Liz's sorrel mare, Babe. They discharged their rifles half a dozen times over Babe's head. Joe decided that discretion was the better part of valour and he lit out for the ranch headquarters, three miles away, for help. Luckily—or unluckily, as the case may be, I was just headed out on horseback to move some cattle to another pasture when Joe arrived. All the ranch vehicles were at work elsewhere on the ranch and so I commandeered a visiting friend, Jim Boydell, and his Ford Bronco, to get me to the scene of the shooting. Joe accurately described to me where the shooting had taken place, but when Jim and I arrived, nothing was to be seen of the hunters, only the brass casings of the spent rifle shells, and a lot of scuff marks that showed where Babe's shod hooves had backed away from the hunters. I tracked the hunters back to their vehicle on the public road, and then lost the tracks. Jim and I ventured out onto the public road, checking to see if we could discover the culprits.

Finally, in an area near where most hunters camped on Grinder Creek, I was waving down cars and trucks and asking questions, when around the corner, one hundred yards away, came an old green Chevy car filled with hunters. They stopped when they saw me and then proceeded forward. I waved them down and stuck my head in the open passenger window. "I am looking for some hunters who shot at one of my cowboys."

To this query came the instant retort, "If that little son of a bitch said we shot at him, he's lying."

Before I left headquarters I had phoned the RCMP and Game Wardens in Clinton, so I knew they would be arriving at the Churn Creek Bridge at the edge of our property about the time that I engaged these trigger-happy hunters. I explained to the hunters that I would like them to accompany me to the Churn Creek Bridge where they could state their case to the RCMP.

They said that they weren't leaving in that direction and drove off. I knew, of course, that there was no road leading off the ranch except the one leading to the Churn Creek Bridge, even if they didn't. The hunters' camp was about half a mile down the road and I caught up with them there. I entered onto their campsite and again requested that they come with me to take their case to the RCMP.

"You can't make us."

"Yes I can," I stated, "if I put you under Citizen's Arrest." I had recently learned about this from a lawyer friend visiting the ranch but I forgot the part that one could do this only if he caught the culprits in the act.

"How do you do that?" the boldest of the brothers questioned me.

"I put my hand on your arm and place you under Citizen's Arrest."

I then proceeded to do just that and was rewarded with a hard punch on the nose and a flood of blood. That's when the fight began. Jim, for his part, when charged by a couple of the men, backed up to his Bronco and reached in the open tailgate to withdraw his rifle which he held by the barrel and swung over his shoulder like a club in an attitude of defence. This action cleared him from any fisticuffs. Meanwhile, I was being entertained by a couple of brothers—as it turned out.

I was sitting on the brother who had punched me, letting my blood cover his face, when the other brother hammered me over the back with a club cut from dead wood, ready for the fire. Things by this time were getting a little out of hand, to put it mildly, so Jim and I departed the scene feeling that we had evened things up a bit. At the ranch house my mother taped up what appeared to be a couple of broken ribs. The upshot of this altercation had these hunters charged with and pleading guilty to a $10 trespass fine in Clinton after being stopped at the bridge by the officials.

We thought that ended the deal until a couple of months later when I received a summons to attend at a Trial for Discovery in Vancouver, after which I was hailed into the Supreme Court of British Columbia to show reasons why I should not pay a large sum of money for assault. Tom Berger was the Steel Workers lawyer, later to become famous for his Mackenzie River Basin Pipeline Environmental Commission and also later Supreme Court Judge of British Columbia. I contacted Tom Braidwood, a lawyer friend I had graduated with from Britannia High School in 1949. Tom Braidwood later became a judge of the Supreme Court of British Columbia, and then an Appeals Court Judge in BC. After he retired he became a one-man commission to investigate the use of Tasers to quell suspected criminals. That was followed up with a one man commission to investigate the death of Robert Djzkansky by Tasers at the Vancouver Airport

All seemed to go well with our trial until the Attorney General of BC refused to allow the RCMP Corporal, who had investigated and retrieved the empty brass shells at the shooting site, to render his evidence. Then we knew we were in trouble. Joe Matson, Jim and Joyce Boydell, and Liz and I spent three days in the Supreme Court of BC, listening to what I reckoned was the biggest pack of lies ever foisted off on a Supreme Court Judge of the Province of BC. I listened to seven hunters describe how Jim had kept them all under rifle subjugation while I, blowing spittle from my mouth like a madman, systematically beat them up. One described how I had "picked him up by the head and he remembered seeing handfuls of his hair wafting away on the wind."

What I learned from this civil trial was that Right does not always prevail, and that seven witnesses beat three in Supreme Court any day. The only saving grace, after I was found guilty of assaulting the seven men, was that the judge decided the trial should never have been held in the Supreme Court (BC's highest trial court) with it's higher scale of damages and should more rightly have been held in County Court in Clinton with a lower scale of damages. Consequently my family paid a sum total of $3,500 in damages instead of $35,000. I would guess that the $3,500—and probably a lot more—went straight to Tom Berger to cover his fee.

Tom Braidwood, for his part, charged me nothing for my defence; I think he felt bad for not getting me off. It wasn't his fault though, for we were up against a stacked deck without the RCMP evidence and it was seven to three in the witness box—actually seven to two, as it turned out. Joe Matson testified to the actual shooting, but unfortunately he was a bit disgruntled and discombobulated by the tactics of big city lawyers and consequently he came off badly in their exchanges. Because of this, the judge said that his testimony should be discounted. .

We had lots of interaction with poachers over the years as they dynamited our metal gates and tore down our fences, but we retaliated as best we could when they parked on our private lands by removing valve stems from their tires and switching spark plug wires on trucks that had not been given permission to hunt. This tactic backfired on yours truly when I flattened the tires of the previous owner of Empire Valley, Henry Koster. He had left his four wheel drive truck parked beside one of our metal gates while he walked in to a good hunting spot that he was familiar with. He hitchhiked to the ranch for dinner at the cook house and it was there that I discovered my mistake. After dinner I had to go down to the Onion Lakes gate and pump up his four big winter tires by hand—a lot of work after a long, busy day.

This particular "smart-alec" trick brought me my "comeuppance" on that particular day.

My dad, Clarence, patrolled the public road leading from Churn Creek to our ranch buildings. He was very fair and if a decent bunch of hunters showed up with appropriate gear, requesting permission to hunt, he would escort them to an area where they could camp near spring water and hunt without fear of being shot by any nearby hunters. Those who had too much to drink or who were obviously ready to shoot up the country would be turned back. Not many of these ruffians took this lying down, but they usually backed down when they tried to intimidate the big rancher.

Dad's final recourse, when they confronted him, was to reach into the back of his pickup truck and pull out a long handled shovel. With the shovel over his shoulder Clarence was as formidable an antagonist as any poacher ever encountered. However that long handled shovel didn't always do the trick.

There was one time I was preparing to head for Williams Lake in our little red 1959 Volkswagen Bug to see Liz, who was eight months pregnant. She was staying with her parents while she awaited the birth of our baby. Dad appeared requesting help facing down some bully boys who had threatened him at Browns Lake. Don Gillis and I followed Dad back to Browns Lake where a group of hunters had just dragged a buck down off the Clyde Mountain, an area within our protected lands. One member of this group was the obvious ringleader and when asked again by Dad to pack up and leave, he started issuing threats.

Dad and Don just grinned at his tirade because they knew they would enjoy the obvious coming fracas. They put up with his taunts for a few minutes until I butted in because I was in a hurry to get to Williams Lake to see my wife. I issued a very strident command for the group to "Pack up and move out." The well-built ring leader, who was quite enraged by this time, turned to me in my "going to town" clothes and sized me up as an "easy go" because Don and Dad had big winter jackets over their large biceps. At 150 pounds soaking wet and a lean six feet, the bully boy decided that I was fair game.

He took a swing at me and was rewarded with some fancy foot work on my part that had him into the lake itself, held upside down and face first into the water. This unexpected turn of events and action by the easy mark prompted him to realize that if he couldn't handle me, he was probably in bigger trouble with either Don or Dad. Whereupon we were treated to all kinds of promises to vacate the property and never come back if I didn't fully dunk him in the frigid waters of Browns Lake. I acquiesced to this and they packed up and left, a very entertaining few minutes for the cowboys, as I recall.

We also had a running battle with Fred Froese and his family. Initially this American family from Woodland, Washington had come to us on very friendly terms to persuade us to allow them to build a road through our private lands into our back country near Red Mountain. The road seemed as though it fit in with our plans for opening up the country beyond Red Mountain to our expanding cattle herds. They proposed building home sites at Roaster Lakes near our Yodel Cow Camp.

Fred, to give him credit, was a master with his ancient D-8 Cat. He would stop the Cat and tear it down if the old girl developed any odd sounds. It took him a good year to build the twenty-eight miles of road from our headquarters through our privately held lands, to Roaster Lakes where the family was to pre-empt land. What they were going to do to make a living never quite became clear to us. However, Fred, his wife, along with his son Jim and his wife Dawn, son Dade and his wife Vi, and daughter Lucille and her husband Jerry had all bought into dreams of wilderness life away from high taxes and the encroaching civilization of Woodland, Washington.

We were on good terms with them for some time, especially big Jerry, who had been a Marine and had lots of stories about the Marine way of life. Jim had also been an MP in the Marines and I think Fred was, as well. They stopped in for a welcoming meal whenever they passed the ranch headquarters.

All went well until Fred ran out of money and began to court any hunters who could pay to get back to where the family had camped, about twelve miles from our headquarters on Crown Land. As a Landed Immigrant, Fred could not get a Guide's license. In addition to this, his hunters had to pass through our privately held lands by way of a locked steel gate situated beside our buildings. Fred became enraged when we would not leave the gate unlocked for his hunters to pass through, even though he knew it was a source of consternation for us because poachers could pass through any time we were not around and especially at night when we were asleep. Thus ended our friendship.

Fred then embarked upon what seemed like an impossible task. He had acquired the most recent map of Empire Valley showing all our Crown Granted (private) lands in yellow. Fred could not build his road through these lots without gaining our permission, which we had given him for his first road from Empire to Roaster Lakes. Somehow, something had escaped all of our attentions regarding the map. In the old days, around 1900, when numerous small ranchers were pre-empting land around the valley, there seemed a necessity that someday a school would be required. One lot at Browns Lake was set aside as a school lot. This lot showed up as green (government land) in the new maps—where our old maps had showed the school house lot in yellow, the same as our seventy-five Crown Granted lots.

Fred got permission from the BC Lands Department to build a new road up the steep rocky hillside behind Browns Lake on the school house lot. If he could get the road built up that steep rocky hillside, he would have access to all the Public Lands beyond. In particular, he could build his new road through Crown Land (land belonging to the government) to where his family was squatting on another piece of Crown Land beyond the Higginbottom Cow Camp. That Higginbottom camp was situated at the western tip of Empire's private lands.

Six miles from our boundary, the Froeses had chosen a squatters' spot that came equipped with a hunting cabin. The hunting cabin on Porcupine Creek had been built by Arrow Transfer's owner, Jack Charles, with Henry Koster's permission. We watched in fascination as Fred made several abortive tries up the steep bed-rocked hill from Browns Lake. Finally, after many days, his persistence paid off and he made it to the top of that daunting hill.

Now the back country that Empire had controlled for seventy years was open to Fred and his family, of course, but also to any other persons seeking entrance to that beautiful, gigantic, wilderness backcountry behind Empire Valley. The Froese Road, as it was called then, eventually became the Black Dome Road after it was extended by gold miners to Black Dome Mountain. There they developed a gold mine that, in its lifetime, was the most productive gold mine in Canada.

The Froese Road became the lightning rod for endless problems for Empire Valley Cattle Company, as outfits like Bill Federation's BC Safaris were able to swing up the new road above Browns Lake and through the Dry Farm—our privately held land—to make camp on Grinder Creek from whence those busloads of hunters could choose to head back into the Empire Valley private lands which were then teeming with deer.

In retrospect, the road was the death knell for the famous Empire Valley bucks. In combination with an undermanned and a bit ignorant BC Game Department, it would take about five years to reduce the once proud herds of mule deer to a minute fraction of their former glory. Perhaps the road itself would not have led to the extinction of the herd, but when the Game Department, in their ignorance, allowed hunters to harvest three mule deer of which two could be does, then the end was nigh.

The herds lost our jealously guarded protection in 1967. When no Canadian buyers could be found after two years on the market, we sold the ranch to Bob Maytag from Colorado. In order that he, as an American, not become involved in the kinds of altercations that he knew we had in the past, Maytag requested that the BC Game Department supervise the hunting in all of Empire Valley. This included the private ranch lands as well as the public lands. Imagine the uninterrupted slaughter of those almost tame does and fawns that had never heard a shot or felt the impact of a bullet until the Game Department allowed unrestricted access in order to kill off this magnificent mule deer herd.

The first November 11th weekend after Maytag took over the ranch—traditionally the biggest weekend for hunters (when Remembrance Day fell on a weekend)—Alfie Higginbottom's daughters were camped at Churn Creek with their families and, for something to do, they counted deer exiting Empire Valley on trucks and cars. In those days bucks were usually proudly displayed with their horns in the air on the backs of vehicles or on the front fenders of cars. On this particular weekend deer were displayed with their four feet in the air indicating the deer were does. That weekend the Higginbottom women counted approximately three hundred does leaving the ranch. What a slaughter.

Today the remnants are picked over every year to provide a few hunters with a very small harvest. It will take a miracle and many closed seasons before the herd will number even a quarter of the approximately five thousand mule deer that once wintered at Empire Valley. On a recent seven-day ride with Warren Menhinnick of Spruce Lake Wilderness Trails, I rode the backcountry around and behind Spruce Lake and the upper alpine areas of the Tyax, Gun and Relay Rivers—the traditional summering areas of the mule deer. I counted but six mule deer bucks on their traditional summer ranges. In its heyday, I would have seen thirty or forty bucks in one day of cowboying.

Don't get me wrong, we had a number of thoughtful, conservation-minded hunters who were good people, out for a mind-refreshing get away in our wilderness area. Many such folks come to mind, but Bruce Ledingham Senior and his sons Bruce and Blair were three of the best. They usually brought with them a few of their executives from Ledingham Construction, which was headquartered in Vancouver.

On one memorable occasion they rolled into Empire on a Sunday afternoon in order to begin hunting on Monday. We guides were worn out, running up and down the steep bunchgrass slopes and gullies, filling out the quotas of our latest party of hunters. Bruce Senior was anxious to get his boys out for a quick look at the country because the boys were so keen, and they had been dreaming about their coming hunt since the previous year. I reluctantly agreed to take them out with brother Dunc and Art Grinder, one of our youngest and best guides,[4] if we could catch a little shut-eye first.

We headed into the Dry Farm area where we knew a fresh bunch of bucks had just come. We intended to show the boys what they had been dreaming about since the previous year's hunt. Instead of just looking, however, Dunc, Art, and I combed a really big herd of recently arrived mule deer out of the three largest gullies, and with the Ledingham party scattered across the flats above, it was much more than we had anticipated. Probably twenty large bucks and fifty does emerged from their afternoon hiding place to be met by the Ledingham sharpshooters. The tally could easily have been ten bucks and every one a trophy. However, the Ledingham crew did not want their hunt to end that quickly so they held their fire after taking only a few. I am eternally grateful for this even though, at the time, this would have been but a fraction of the huge number of bucks around. Besides, what would the Ledinghams have done with their time other than enjoy the scenery for the next three days? Considering that their three day hunt was still ahead of them, I guess they would have spent those remaining days doing little more than *armed hiking*—to use an expression coined by my son-in-law, Paul Bucci.

My buddy, Nick Kalyk, who used to help out at haying and cowboying time, introduced us to his uncle—another Nick Kalyk—from Burnaby. Uncle Nick would come with his pals, George Gracious and Peter Paulus. Uncle Nick loved to take a horse by himself and head out to the Point Pasture, an area above the Fraser River. For two days in a row he would return from his hunt, swearing that there was an elk running with the other bucks and does at the Point. I insisted that there were no elk on the ranch. On the third day of his hunt, Nick talked me into taking his party by four wheel drive out to see this elk. Of course, the "elk", which Nick then killed, was the largest bodied buck that I had ever seen, with antlers so wide that they barely fit into the back of the pick-up box. I guestimated that, with head and horns on, it would have weighed about 450 plus pounds—it was all that the four of us could do to heave it into the bed of the pick-up.

The Kalyk buck was never weighed, but Dad and his pal, Bill Richardson, on one of their morning poacher runs, had spied another huge buck in the Gap hayfields on three or four successive morning jaunts. Feeling that a poacher would end up with this classic four-pointer, because he often grazed near our main road (which had public access) they decided that Dad should kill and put his tag on the freshly killed monster of a buck. Bill took the field dressed buck (gutted with legs removed at the knees) back to Vancouver with him where he had the buck cut up, wrapped, and frozen. The buck weighed three hundred and seventy-five pounds in Vancouver. Bill had the buck's head mounted for Dad to hang in the main ranch house. The measurements of the head would have placed it high in the Boone and Crocket Club List of trophy bucks in North America. This was the only buck that Dad ever had mounted. The trophy head still hangs on the wall of his grandson, John Gillis, at his ranch house in Stavely, Alberta.

John Fraser, a timber cruiser and part owner of Cassidair Services in Nanaimo, killed a buck with the widest horns I had ever seen. That buck's horns would not fit sideways in a pickup box that could hold four by eight foot pieces of plywood. That's very big! John was a wonderful guy, and on one of his trips to Empire, he brought his airplane with him and showed me how to fly. After that I was hooked on flying. John was unfortunately killed flying with a Vancouver Island log shipper when he tipped a wing tip into a floating ocean log boom that the log shipper was having him survey from the air. I presume the fellow's eye sight was not good and that he had John get close enough to the logs that he could pick out his company's unique stamp on the ends of the logs.

Probably the biggest buck we ever saw at the Empire was never harvested. Dunc and I would spot this granddaddy staring down at us from the highest slopes above Shakes Gulch on our way back from successful hunts of the Big Churn Creek Country—where our cow herd was wintered until the snow got too deep. We often tried for him, but he was as ghostly as he was big. To give a seasoned hunter an idea how big he was, if one looked straight at him through a four-powered scope, the base of his horns bisected the distance between his hooves and the tip of his horns. On a granddaddy like that it would mean that the height of those antlers would have been in the forty inch plus range. Sometimes the bucks win.

As a guide, I had my share of difficult incidents, but probably the worst was perpetrated on me by an ex-sergeant of the US Marine Corps (now a landed immigrant). He always wanted to hunt by himself on Old Whitey, the same mare that packed Francis Haller home from Yodel Camp when he had his heart attack.

The sergeant practised dismounting from Whitey to ram a fistful of cartridges into the clip of his rifle. He bragged that he could dismount, load up, and shoot within five seconds, and watching him practice, I had no doubts. One particular fall day he arranged to meet us at The Point, near Hog Lake. We had a fairly large party with several guides and I mixed up a pretty good hot lunch over the fire when we met the sergeant. He reported that he had filled up his quota of bucks while on the way to meet us. I told him, in that case, he would have to haul in his bucks and keep the fire going until we finished our hunt, at which point we would pick up his bucks for him. He was leaving the next morning for a logging show at the Nhatlatch Valley across the Fraser River from Boston Bar.

When we arrived back at the fire, the aforementioned sergeant was nowhere to be seen, though his bucks lay by the side of our four wheel drive road. Imagine my consternation and dismay when, before leaving the next morning, he took me aside with something of a hangdog look to his face.

"Uh Mack, I have bit of a confession to make to you." he said.

"Uh oh. Alright let's hear it," I said. "What happened?"

"Well Mack," he said, "I just couldn't help myself. After you left, I went over to the edge of the bluff and looked down and counted seven bucks ranged across the side-hill below me. I hate to say this but I picked off four of them and spent the whole afternoon getting to them, gutting them, and covering them with sagebrush, so the eagles wouldn't eat them."

I will refrain from relating the remaining conversation verbatim. Needless to say, the ex-sergeant and Empire Valley parted company for good. I told him in no uncertain terms that he was not welcome at Empire Valley from that day forward. Then I spent that whole day with pack horses bringing in the bucks. Finding them, even though his directions were very good, took me half the morning because he had covered them so well. Loading and tying them onto the pack saddled horses on that slippery slope was difficult enough without me sliding under the horses' bellies on the frozen ground. That was a really tough job.

It took me the rest of the day to load and get the bucks home and onto the deer hitching rack. Fortunately there were hunters who had not filled out their quotas, who relented and took care of these extra bucks upon leaving. These hunters' actions took me out of a legal and ethical jam and for that I was very grateful. We could have lost our guide's licences for killing more than the allotted number of deer per hunter. As I recall, John Wright from Nobody's Stores in New Westminster and his friend Jack Wiley, a North American Skeet Shot Champion and their hunting companion, Glen (name is lost), were good enough to take the extra bucks off my hands to fill out their quota.

There were so many good folks and good memories associated with hunting at Empire Valley. I really feel quite blessed, looking back. Bill Richardson and his wife Pearl came every year to hunt, to help out with the cooking and with poacher control. Bill often brought one of his Pioneer chain saws that his company manufactured in Vancouver. He also custom built his own beautiful hand crafted rifles. He built one for Dad, one for Don Gillis and one for me. I lost mine later in the Fraser River in a disastrous jet boat sinking.

Some of our other favourite hunters have since lost the urge to hunt, but we have not forgotten Doug (now deceased) and Molly Beaumont from Victoria who would pitch in and help with the ranch work, as well as hunt. Molly usually helped my mother, Eleanore, if she was called upon to cook if we had lost a cook. Alice Haddock from Pender Harbour would also help when she came with her husband Johnny who would go patrolling for poachers with my dad.

We ran a large cook house which seated approximately thirty people—hunters, guides, and ranch workers—at two long tables. Once in a while a cook would be overwhelmed by the work and would quit, so Mother and my sister, Donna, would have to do double duty. We did have some wonderful cooks who could produce great meals for hungry hunters, as well as the ranch hands. Half a dozen passersby, stopping in for dinner, were always welcome and never phased the likes of Clara Paulson, Bessie Zimmerly or Marion Eichart among others. Marion asked to come back every year at hunting season as she loved the whole atmosphere.

When I think of those fabulous meals, I remember one November day in 1963 in particular. We were all gathered for lunch; my dad waited till we were all seated before he solemnly stated:

"I regret to inform you that President John Kennedy has been shot and is not expected to live."

I shall remember to my dying day that this did not elicit sorrow from the four US hunters sitting at the table, all true Republicans, of course. Liz, on the other hand, I found sobbing uncontrollably when I returned to our own little house after lunch. Politics is a strange game and I know that it has no place at the hunters' table, but I still can't get out of my craw those hunters' strange reception at hearing of the death of their president.

—

[4] Art was later killed in a head on crash with an ore truck on a wintery, slippery road, while driving with friends on their way to work from Ashcroft to Bethlehem Copper Mine in the Highland Valley.

Chapter Fifteen
Bear Bait

Francis Haller arrived back at Empire Valley Ranch from the Tyax Country one fall trailing a tall young guy from Gold Bridge by the name of Don Tremblay. (Don is featured in my Thanksgiving story.) Francis thought we should hire the kid on and the fact that Don had taken a farriers course down in Montana made him a useful addition to our cowboy crew. We were always running about fifty to sixty head of saddle horses that required frequent shoeing on account of the rocky roads and trail that led into our back country.

My job was to hard-surface all the relatively soft iron horseshoes with a copper welding rod to extend their useful life. Hard-surfaced shoes could stand any amount of rocky trails while iron shoes would wear out quickly. I think of myself as a pretty fair welder even though my uncle Duffy used to suggest that a welder with the shakes of a hangover makes for a better welder. Be that as it may, my hard-surfaced shoes lasted for many re-shoeings. Of course, that didn't prevent the horses from sometimes stepping on their heels in the mud to pull off the shoes because of a hurried horse shoeing job—I'm sure my horseshoes kept sharp-eyed horsemen on Spruce Lake trails supplied for years.

Don Tremblay was a good farrier. There was no arguing that, but he had a bad habit of banging his shoeing hammer on the horse's foreleg if it persisted in pulling its foot away while he was shoeing. It was that while re-shoeing Francis' top horse, Spooks, that Don cracked a small bone in Spooks' foreleg while giving her his customary rebuke. After Don's misadventure with Spooks at the Spruce Lake camp, she was turned loose in hopes that she would heal up, but the injury only worsened and she was permanently lame. For their parts Francis and Dunc, who both had an interest in Spooks, were heart-broken, as Spooks could not be replaced.

Sometime later, we were camped at the Relay Creek Camp, which we had inherited from Rube Fast and his Magnet Mine crew—they had been trying to develop a mercury deposit on the south side of the Tyax River—when Joe Bingham showed up, looking for any bear bait horses we might have.

Joe was a one-time longshoreman, a huge man with a great bellowing laugh. He wore a big black hat pulled low over his eyes and often had a six-inch stogie jammed in his face. Joe had parlayed a big salary into the purchase of a broken down lodge on Tyax Lake and was guiding groups of back-country enthusiasts. At that time Joe was struggling to make a go of it. We wintered his horse string at Empire Valley to give him a little break from buying expensive hay which had to be trucked to Tyax Lake. This not only saved him money but provided us with a bit of cash flow as we cared for his twenty or thirty horses.

I liked Joe, so when he came looking for bear bait, I decided that the time had come to put Spooks out of her misery. There was no room for a permanently crippled horse at the ranch—top horse or not.

Joe and I saddled up one morning and rode the twenty miles over to the Spruce Lake range where we ranged out any spare horses; I knew Spooks was with them. When we arrived I discovered that an albino mare that Jim Davies had given us—she was of no use to him and, it turned out, was useless as a cow horse as well—had been attacked by a grizzly who had raked her on both sides of the withers with four inch claws. The grizzly had obviously not been able to hold her and she escaped back to the herd of spare horses. By the time we got there, her injuries had festered, and maggots were crawling out of her wounds. I told Joe that she could go as well.

We caught the two mares and led them back towards Relay Creek. Along the way we led Spooks up onto the New Range, in the alpine country above Bannock Camp and shot her in a spot that would allow Joe's hunters to hide behind a spruce hedge while waiting for a hungry grizzly. When we got back to the Relay Camp I advised Joe to take the albino mare further along to the Mud Lake Camp where I had most recently seen grizzly tracks. Joe continued on the ten miles to Mud Lake where he shot the mare and then he returned to Relay Creek. He had originally intended to stay overnight with us, but instead he decided to drive into Gold Bridge and pick up his hunters so that they could be back to the New Range horse kill at dawn in case a hungry grizzly arrived that soon.

The next morning he left our camp with his hunters at about 4:30 AM and they rode the twelve miles to the bear bait on the New Range.

As a side note, baiting grizzlies in this fashion was legal then, but this practice has passed into history. The unfortunate upshot of this is that the area is now overrun with grizzlies. Not long ago my friend Barry Menhinnick had barbecued some steaks at his log home on Gun Creek. He had forgotten one steak on the grill. Hearing some rattling noise out on his deck after dark, Barry turned on his deck light and looked out through the window on the door, only to find a full grown grizzly standing and staring him in the face. That close-up sighting did give him a bit of a pause. A lot of us might have lost control of our bladders at that scary sight.

Back to Joe and his hunters—imagine their consternation when daylight arrived and they were in position to ambush a hungry grizzly, only to find that the dead horse was no longer in view. After a half hour wait, they rode down to investigate. They followed the tracks which showed that a large grizzly had dragged the carcass down into a patch of timber below where the horse had been shot. All that remained of the carcass was the heavy frontal portion of the horse's skull and four hooves. The tracks indicated that one grizzly had eaten that 900 pound mare in one sitting—a few hours. That gargantuan bear feast certainly does give anyone pause to reflect on the voracious appetite of that obviously huge, hungry bruin. Not a bear that you and I would want to have a face to face meeting with.

Joe and his hunters camped out at Mud Lake for a week but no bear approached the albino mare's carcass—I guess grizzlies just don't like albinos. So ended Don Tremblay's slip of the shoeing hammer. I know that even though he was an excellent cowboy who took over Francis's job managing our cattle in the backcountry when Francis was hospitalized, neither my brother Dunc, the actual owner of Spooks—nor Francis—ever forgave Don for that miscue, which cost them an extremely valuable and much loved cow horse.

Don Tremblay had quite a scare managing our cattle once. He was moving about thirty head of heifers up and over the Gray

Rock promontory in the Tyax Valley when his lead heifer was suddenly seized by a huge grizzly. The cattle and his pack horse bolted back down the trail, and Don refused to use that route any more. Instead he would detour around the Tyax and up Gun Creek to Spruce Lake, a thirty mile detour! All in a day's work, I guess.

Chapter Sixteen
Incidents

A great deal of one's time on a ranch is taken up by routine, be it daily, monthly, yearly, what have you. It is never boring though, for the sudden vagaries of ranch life cannot be ignored.

Early in our tenure at Empire we decided to switch our main ranch house cook stove over to propane from wood. To do this we needed a really big propane tank to be brought in from Clinton as there could be periods of three to four months when a propane truck could not get through to the ranch due to impassable roads or bridges which might be out. We ordered Mother a new four-burner propane stove for the big house so that she could cook in the winter when the cookhouse was shut down. A women's work is never done.

The five thousand gallon tank was about twenty feet long, so Dunc headed out with the old cab-over Austin cattle truck to Clinton, where he picked up the tank filled with propane gas. Dunc managed to get the twenty-foot tank onto the sixteen-foot deck of the Austin, but obviously there was a four-foot overhang. With the consequent transfer of weight behind the rear wheels, the Austin's steering was, to say the least, sluggish and unpredictable. The trip went reasonably well as Dunc nursed the load out from Clinton on the relatively flat, though twisted roadway, leading to the Fraser River.

On leaving the Fraser the road climbs, quite precipitously at times, on the way to our headquarters. Dunc got as far as Little Red Mountain before he got into trouble on a gear change on that hill. This stopped the truck's forward motion momentarily until he completed the change. Then the truck jerked ahead causing the tank to slip back another couple of feet. Because of the shift in weight, the truck could no longer manoeuvre at all, so Dunc walked home to get a tractor with a front-end loader in order to push the tank forward again. Don Gillis and I joined him to complete his journey.

After repositioning the propane tank with the aid of the tractor, Dunc announced that he "wasn't going to drive that bomb any farther." So it fell to me to complete the job. I got the old Austin headed uphill and in short order managed third gear in the low range so as to get a run up the hill. I had only gotten halfway up the hill when Dunc passed me on the fly with those long legs of his, going flat out. I could see the panicked look on his face through my open window.

"Get out, she's going to blow!"

Fortunately for me, this was not the case, for I took a few minutes to get the old girl shut down so that I could discover the problem. It seems that, in bouncing the old beast up the hill, I had upset a jerry can of water on the tail end of the truck, which Dunc had kept there in case the old Austin overheated its radiator, which it often did. The resultant stream of water had convinced Dunc and Don, trailing me in the yellow Chevy pickup, that the propane tank had been damaged with my jouncing to and fro. If it were true then that much propane would be more than enough to blow me and the Austin to Kingdom Come.

I did appreciate that Dunc risked his own life to warn me of the apparent explosion and all is well that ends well. The tank did make it home and my mother cooked some memorable meals on her new propane stove. I can still see Dunc pumping those long legs of his and hollering as I tried to herd that recalcitrant Austin up the Little Red Mountain Hill.

The old Austin cattle truck was certainly a menace, and a couple of times—once with Dad driving and once with me—we missed gears going up hills leading out of Empire to Clinton, while loaded with cattle destined for the railhead at Ashcroft. Both times cattle were thrown backward, through the tailgate, and out onto the roadway. That meant we had to go home, get a saddle horse, round up the cattle, find a loading chute and reload the anguished animals. This was all in a day's work for cowboys, who weren't cut out to be truckers.

Another time I had hauled a load of dry cows to Kamloops with the old beast in the dead of winter. Mother had prevailed on me to ship some cattle, for she had no money to pay our bills. She was our bookkeeper as well as our winter cook. Apart from the horrendously bad prices I got for the dry cows, all went well until the return journey with the empty truck. I slid off the road at the notorious Coal Pit section of our Empire road leading from Churn Creek to the headquarters.

I had been attempting to cut to the outside of some really bad drifted snow blowing down off the high banks of the Coal Pit. Even with chains on I couldn't get the old beast going forward again so I got out the snow shovel and attempted to clear the snow out from under the rear dual wheels. The wheels seemed to be almost off this very narrow section of road at this point. I was not able to clear out enough snow to reach the underlying dirt road but I was continuing to shovel when Dad came around the corner with the TD-9 Cat.

I was relieved and happy to hear the sound of the Cat's diesel motor and appreciated Dad's foresight in coming to get me. I was not, however, prepared to find that once Dad had hooked onto me with his winch, the truck almost tipped over the bluff. What I had thought to be ground under my left rear wheels was just hard-packed drifted snow. Had I continued digging with my snow shovel, I might have had the old Austin cattle racks tipped onto my back, with little warning. I disliked the old beast enough already that I certainly didn't want her lying on top of me while we slid into the Fraser River below.

Vehicles and I didn't really get along. I was much happier on a horse, but cars and trucks are a necessity when you live seventy-five miles from the nearest town and general store. I think the best of those vehicles was Liz's and my '59 Volkswagen Beetle. The fire engine red car had been given to us as a wedding present by Liz's parents, Anne and Doug Stevenson. It had been specially ordered for Anne a year or so previously and she particularly wanted the fire engine red colour that was not part of the regular selections offered by the dealership. That car subsequently made believers out of dozens of Caribooites who bought Volkswagens from the dealer who had sold ours in Williams Lake, based very much on our Bug's reputation.

One time Liz and I took our baby Lisa to Williams Lake where, in the company of the Stevenson grandparents, she would be christened. The road was a snowy hell, with drifted snow particularly bad across the section called the Dog Creek Airport Road. En route Liz had to drive many times, much to Baby Lisa's delight, while I dug a narrow path through three or four feet of drifted snow. At other times we could drive on top of the packed drifts and we used the tops of the snow covered fence posts alongside the roadway to guide us. At one point we passed a Public Works Department grader, with chains on all four wheels—it was stuck in the drifted snow.

Winter Roads at Empire; Dunc on TD-9—Liz in pickup truck

It took us almost six hours to reach Williams Lake, a trip that took two and a half hours during good weather. We were quite pleased with ourselves that night at the Stevenson household—sitting around a welcome fireplace, with welcoming parents, drinking very welcome hot rum—to hear Doug Stevenson report that Tommy Desmond, the Cow Boss of Circle S Cattle Company at Dog Creek, had come into Mackenzies Ltd. for supplies. Tommy told Doug that he had followed us into Williams Lake with two three-quarter ton four wheel drives loaded with salt for traction and chained together. They would plow one truck as far as it would go into the drifts and the other would pull it back out. That whole process took them seven hours from Dog Creek, to our six from Empire. Goes to show that a good man (me) on a snow shovel, a hard driving wife (Liz) and a happy baby (Lisa) beat those mechanized contraptions any snowy day, when the hot rums beckon. Well, Lisa didn't get any rum—more likely, warm milk.

Speaking of new contraptions—the first time I thought of using an airplane to look for lost cows, I remembered that Henry Koster had a friend named Les Kerr who operated Skyway Air Services out of Langley. I contacted Les and he obliged me by bringing in a Piper Super Cub to look for cows. One particular incident stands out in my mind. We had returned from spotting cows on the wind-swept, snow-covered slopes of the backcountry and had found several lost cows. I mentally noted where I could find these later when I went in on horseback. Les had flown us home after one of our flights into the backcountry. When he arrived at a point at the edge of a bluff a thousand feet above the cook house, he dropped the Super Cub into a dive and pulled the plane up just above the cookhouse chimney. The result was that the down draft caused stove soot to billow out over our cook, Clara Paulson, who was just bending over to check on the dinner. On looking back, with a waggle to his wings, Les and I could see a blackened Clara Paulson waving her fist at us from the cookhouse door. To give Les credit, he would not have wanted to cover Clara with soot, for Les, as much as anyone, enjoyed Clara's wonderful cooking.

The last time I used the services of Les Kerr[5] and his Super Cub was in 1963. We took a last flight over the backcountry to discover six lost heifers on the upper slopes of the Tyax valley near Relay Creek. When I got home, I quickly loaded up a couple of sharp shod horses, some hay and oats for the horses, and some grub for me. It was getting to be late November, and if I was going to save the heifers, I would have to get cracking. All went well as I trucked the horses from Empire to Lillooet and up the Bridge River Highway to Tyax Lake.

Suddenly a big weather change occurred and some heavy snow started as I left Tyax Lake bound for the Manitou Mine site. Soon there were four inches of snow. As I got closer to the Manitou Mine and Relay Creek, the snow had piled up to almost a foot. I put on chains and ploughed along, but the last six miles of road to the Manitou site were built over a section of rocky hillside. The road builders had been forced to slope the road at quite an angle to get through. When I reached this part of the road the truck started sliding towards the creek a hundred feet below. I had only two single tire chains so I put them both on the dual tires on the upper side of the road. In this manner I was able to get about two miles from the Manitou, where it became just too tough to continue, so I packed it in alongside an old dilapidated mine building.

I unloaded my horses and set up camp there in those primitive conditions. The next day I rode up into the Tyax Valley to find if the heifers were still in the vicinity of where I had had discovered them them from the air. They were not far from that area but when I found them they were quite wild. I managed to run them onto our cow trail and then out of the Tyax and down to Relay Creek. The river was frozen over with glare ice and the heifers refused to cross. I took off my cowboy hat and used it to pack sand and gravel from a nearby exposed bank to sand a trail across the glare ice. That took about three quarters of an hour, and by that time the heifers had disappeared into the brush along the river. I located them and drove them back to my hand-made crossing. Four of the six went across but two got away. Figuring four out of six is better than nothing I drove those four up the hill out of the Relay Valley and four or five miles on our road towards Mud Lake in the direction of home.

That night it got very cold so I took my horses inside the old building to get out of the wind and snow. I fed the horses some hay and oats and made myself a meal. The horses appeared to enjoy their bivouac but sleep for me was not easy as the strong smell of horse urine was sleep-depriving. The next day I located the two missing heifers and got them across my gravelled path. They were even wilder than before and once across, they split up and I lost one going up the side-hill out of Relay. So I spent another night of urine smells after running the one heifer a few miles after her buddies. The next day I found the last wild heifer high up on the Relay Creek side-hill. I was riding Duke and we high-tailed it in behind the running heifer to drive her down towards the cow-trail. It was a beautiful cold sunny day and as we raced down the side-hill, we were alternating between sun and shade under the park-like forest canopy.

Suddenly I was forced out of the saddle by a dry dead snag. I thought that I had lost an eye, for that was where the snag had caught me. From my side, as I lay on the ground, I packed snow into what I thought was my eye cavity and when that filled with blood, I packed more snow till the bleeding stopped. Duke was waiting patiently so I climbed back in the saddle and headed for camp. I consigned that crazy heifer to purgatory or the grizzly bears, whichever came first. With the aid of the truck's rear view mirror I discovered a huge cut under my eye, but thankfully the eyeball was still intact.

Next I cut some wood for the fire but because of the poor visibility from my eye and the descending darkness, I chopped a big chunk of flesh out of the palm of my hand—the scar is still visible today. Again I packed snow into the wound to stop the bleeding. The next day I followed my little lost heifer to determine that she was following the Cat road and cattle tracks to catch up to her buddies. Then I loaded up Duke and Rusty, changed my chains to the opposite side of the truck's dual wheels and headed for home. This time all went well, and we arrived home safely except for a few cuts and bruises to the cowboy. That little incident makes me wonder what I would have done if I had really hurt myself, twenty five miles from the nearest habitation.

I have mentioned the "switch blade horns" of the cows that we purchased with the ranch from Henry Koster. This little anecdote portrays what could happen to a cowboy in a "face to face" confrontation with one of those dangerous cows. Don and I had driven such a cow into the Bishop Ranch stackyard to work on her prolapsed uterus. This condition is caused by the cow pushing a large portion of her uterus outside the vagina, long before calving time. This operation required that we rope her and lay her on her side so that I could push the uterus back into her innards and then staple the vagina closed with special wooden ended steel rods that I had purchased from the vet in Kamloops for that purpose. It seemed that we were getting several of these problems. Evidently, when the cow eventually calved after we removed the staples, if the resulting offspring were a heifer, she would inherit the problem.

This time Don and I were both riding colts as we roped the cow's front and rear legs. Don stepped off his horse to hobble the cow's rear legs so I could work on her. Unfortunately his horse stepped forward enough to loosen the rope stretching from the cow's hind legs to the saddle horn. The cow immediately jumped up and pinned Don to the side of the baled hay stack. My horse Shifty held the straining cow for all she was worth as the cow swung her horns back and forth in front of Don's face. Gradually Shifty pulled the dangerous cow with those dagger horns away from Don so that he was able to save himself. "Whew," Don gasped, as he resumed breathing, "That was awful damn close." We resumed work on the prolapsed uterus, with the cow laid out flat and Don staying on his horse this time, until I hobbled the cow. After finishing my job I turned to look at Don and noticed for the first time that a drop of blood was on the end of his nose. "You don't know how close that was, Don. Those horns almost wiped your nose right off your face."

One sad incident comes to mind. I was at the home ranch when a Department of Transport Aeronautical Investigator and an insurance man arrived requesting me to take them out to an area near our Yodel Camp. I recognized from their description that the site they were looking for was in Starvation Canyon. I drove them and their gear in a ranch 4X4 to Yodel Camp and then pioneered the truck to the crash site of a single engine plane. It seems that a Mooney Mark 20 A- plane had left Vancouver in poor weather to fly to Quesnel. On board were the pilot, his brother, and their wives. Evidently they owned the Patchett Brothers Sawmill in Quesnel. It seems that the plane must have run into a fierce snowstorm over Red Mountain, the pilot lost contact with his horizon and the plane flew vertically into the swampy ground of Starvation Canyon at approximately 160 mph. Luckily a Search and Rescue Buffalo Aircraft passed almost directly over the site and a vigilant spotter could see the tail of the plane sticking out of the swamp. Once the plane was found the DOT sent a helicopter in, only to discover that no one had survived. They were able to come in to remove the bodies before we arrived. As it turned out, the plane had only clipped the limb off one jack pine tree as it plunged into the ground. After surveying the scene, the inspectors had me dig out all the mud around the engine. They needed to examine the plane's propeller to find out if it was nicked from flying at full power into the ground. If there was no damage, then the indications would be that the motor was not functioning on impact. Using this method, we discovered that the plane had been under full power and, as well, the airspeed indicator was locked on at top speed. As the insurance company was prepared to write the aircraft off, I asked if I could retrieve the motor. After some hard work on my part, I detached the motor and we loaded it on to the four wheeler. Later I took the motor to the airport in Vancouver and located a company that said that if I gave them a day to subject the motor to a flux test of the cylinders, and if all was well, they would buy it from me. The next day they told me that the engine was just fine and so I pocketed $350.00, which was a lot of money in those days.

In 1964 we at Empire played host to several members of the Grey Cup Champion BC Lions. "The Lions Roar in 64," if you remember the saying. I had come down from my house after breakfast, to go to the cook house to meet with the crew. Along the way I stopped off at my parents' place and was unexpectedly introduced to several big football players. I well remember greeting Mike Cassic whose gigantic hands made mine feel as though they belonged to a child—and I have big hands. I had gone to Britannia High School with his older brother Frank, who was the fullback on our school's football team.

After breakfast we took some of the boys out for a deer hunt, which is why they were there. Art Grinder and I took Willie Fleming and Bill Muncie up to Onion Springs, near Hog Lake, for our hunt. I had Art take Willie, and I guided Bill. My intention was to walk these men down a gentle incline of the rangeland to meet at the old Bryson Cabin at the bottom of this hill. We had not gone far before I could hear this terrible wailing coming from Art's direction. I put my field glasses on the pair and then called over to Art.

"What's the matter?"

"Willie says he can't walk downhill," hollered Art.

"Then pack him." I hollered back and proceeded down the edge of a ravine trying to scare out a buck. When I sensed Bill had stopped behind me, I turned around and asked him what the matter was?

"I can't go no further," Bill said in a strong southern accent.

So after a short rest, I ended up, like Art, shouldering a football player down the rest of the hill. It turns out that these fellows could run 40 yards in under 5 seconds on a flat field but their leg muscles could not adjust to going downhill. As I recall, Bill Muncie was instrumental in the BC Lions' victory over the Hamilton Tiger Cats that year by scoring two touchdowns, one the winning touchdown. Willie Fleming scored one touchdown. Score: 34-24 for the Lions.

We, at Empire, had not been able to watch the game, for we had no television on the ranch. I had a Kamloops TV station technician come to the ranch and we tried putting a receiver in many different places but we had no luck with any kind of reception. Not long after we sold the ranch to Maytag, he was able to bring in television, something that would have made life a lot less lonely and a lot more bearable for Liz.

Other Lions players who arrived that day were Neil Beaumont, the Lions' fullback, and Steve Cotter, a Lions' offensive guard. Steve and I wrestled in the ranch yard for the entertainment of the crew. He was big, strong and easy going. He and his family eventually moved to Kamloops where Liz taught his daughter Jennifer in Grade One at Dallas Elementary School. She became friends with Steve's wife Sharon, also a teacher. Liz and I also corresponded with Neil and his wife for a while, and they welcomed us to dinner at their home in North Vancouver.

To a man, that group of football players was a great bunch of guys and they well deserved to win Lord Grey's Cup that year.

[5] Les Kerr went on to start a big aircraft company that had most of the contracts for fighting forest fires in Canada, with water bombers stationed in Abbotsford.

Chapter Seventeen
Clarence Bryson: Larger than Life and Revered by His Family & Friends

Clarence George Bryson was one heck of a man. He was born big, grew bigger, and was larger than life in his trademark white Stetson cowboy hats. During the depression he took on the challenging task of milking Hereford range cows to sell their milk to pay taxes on the family's Grange Ranch at Pavilion BC.

Throughout the coldest of winters he used a bobsleigh to haul mining parts and supplies down off Pavilion Mountain to the Pioneer Mine, located far below the Pacific Great Eastern Railway tracks (now called BC Rail). To accomplish that difficult task he used a single harnessed horse to pull the sled down the single track trail, which was down a scary and precipitous incline to the mine, situated on the banks of the Fraser River. To keep the load from sliding off the trail into the Fraser River below, Clarence ran alongside the sleigh and bolstered the load. The harsh weather made for difficult working conditions, even for a fit young man in his twenties, especially because he did not wear any protection from the cold on his hands or his ears. In fact, I never knew my dad to use any head gear to keep warm, other than a cowboy hat, no matter how low the temperature dipped.

In 1949, when I was 16, he and I fed 500 head of Nicola Stock Farm cows on our newly purchased Voght Ranch in Merritt, BC. The cattle were driven from Nicola to Merritt on an old, unused connecting railroad. The Cow Boss of that short cattle drive was Slim Dorin, who became a close friend of the family. He was assisted by another cowboy named Blondie Ellingson and I remember they arrived wearing only ear muffs over their ears in that minus 35 degrees Fahrenheit weather. The average temperature of that winter was between 25 to 35 degrees below zero. My mother managed to thaw the cowboys out with homemade soup and hot coffee.

We fed the Nicola Stock Farm cattle loose hay with pitchforks. I would climb on top of the stack and pitch-fork hay down to my father who was loading the sleigh with the hay that I pitched down for him. We had converted a horse drawn sleigh to be pulled by our tractor. At times, while off-loading the hay to the waiting hungry cattle, I would put my pitchfork handle into the exhaust pipe of the tractor to warm up the wood because, even though I was wearing gloves, my hands felt as if they were frozen. That was the only time I ever saw my dad cover his ears with his bare hands, to warm his ears. I used to say "no sense—no feeling" to explain his cold-hardiness and to tease him. The whole scenario still makes me shiver, just remembering it.

Dad on rock–with Mack on horse –starting a cattle drive

The first day on our new Merritt ranch, we were not even settled in when we could hear fire sirens blaring. We looked to see one of our haystacks was on fire. I stumbled across dark barren fields and fell into ditches trying to get to the fire. A lot of the town's young people came to help the volunteer firemen put out the fire. Somebody produced a hay knife and a great strong young guy named Eddie Annheliger and I sawed the stack in two, leaving the smaller end to burn itself out. That is how I met my future Merritt buddies Ron Cressy, Joe Laviguer, Ray Emmerick, and Eddie.

Clarence was born on the Grange Ranch in Pavilion in 1910. He was the third son of Minnie and JB Bryson in a family of five boys and one girl. Clarence and his younger brother Duffy managed the ranch when they and their siblings inherited it from JB Bryson upon his death. After Duffy came back from World W II, Clarence managed the mountain section of the ranch, while Duffy managed the valley section. Clarence and his brothers had built a log cabin on the mountain portion of the Grange Ranch in Pavilion. It was used as a bunkhouse when they were working on the mountain. When Clarence began managing the Mountain Ranch he built a frame house that our family moved into in the nineteen forties. The log cabin was where Dunc and I stayed and when our Vancouver cousins came up in the summer all the boys would bunk together in the log cabin. After Donna and I began staying with the Brett family in Vancouver during the school year, the kids would all come to Pavilion for the summer to help with the haying and cooking.

During the war my dad and my grandfather Donald McCallum, my mother's father from Vancouver, built a frame house for our family in the valley below Pavilion Mountain. The new house was quite near to my grandfather JB Bryson's eleven-bedroom ranch house. JB's house, as well as Robert Carson's house (previously mentioned), served as a stopping house on the old Cariboo Road. The stopping house was at 25 Mile, measured as being 25 miles from Lillooet, which was Mile 0 on the Cariboo Gold Rush Trail. That was the reason that my family's livestock brand was the number 25, branded on our cows on the right hip and the horses on the left shoulder (hence Two Bits the bucking horse).

In September of 1937 Donna started Grade One at our little one-room Pavilion Elementary School. In December of that year the school was in danger of closing because a family had moved away, leaving the school one pupil short of the number of children required in order to keep a teacher. The other parents approached my father and mother to see if there was any chance that "little Mackie" could start Grade One in January. I was not going to be five until February, but when the school reopened after the Christmas break, with my addition, there were now enough students to keep the school going. When the Inspector of Schools for the Lillooet School District heard that a four year old was attending school, he made plans to visit Pavilion Elementary to send me home. The winter driving conditions that year made the road impassable and so his trip was postponed until after spring breakup. When he did arrive he found—according to my mother—that I was the best reader in the class and so he allowed me to stay. I went on to Grade Two with Donna the following September, thus starting a school career where I was to be two years younger and much smaller than my classmates. We had no vehicle so Donna and I would walk the three miles to school everyday—uphill both ways, as I tell everyone.

We converted to riding horses when we moved to the Mountain Ranch. My sister Donna, brother Dunc and I rode down the mountain on a steep trail to the valley and then a further three miles to our one-room school at Pavilion. In the school were some of our Carson second cousins—notably Pat and Jack—who also lived in the area. Dunc used to get Donna and me quite angry with him because he would gallop his horse Pinto down the fairly steep mountain, often short-cutting straight down the slope rather than following the switch-back trail. Donna lost her horse that she called "Plug Without a Spark," because one day she tied the mare up with enough slack in the halter shank to enable the mare to eat some grass. She thought she was doing the horse a favour but it got tangled up in the rope and choked to death

My father's brother Duffy had been an aircraft mechanic during the war, as was their youngest brother Glen. Glen was stationed at an air force base in England. When he came home from the war he brought home his war bride Hilda, who he met and married in England, as well as his baby daughter, Glenis. Duffy was stationed in Winnipeg, servicing the war planes headed for England. Their oldest brother Robin fought his way up the Italian Boot with the Canadian Forestry Corps, building Bailey Bridges ahead of the advancing Canadian Army. The Nazis had blown up their existing bridges behind them as they retreated. Norval, the second oldest brother was needed at home to manage the Public Works Department's yards at Lillooet, servicing the area's hundreds of miles of gravel roads.

Dad tried to enlist but they would not let him because he was needed to run the family ranch with his father, JB. Ranchers and farmers were seen to be part of the contingency needed at home to keep the agricultural sector in production. Dad was disappointed not to be going off to fight for his country. Norma, the youngest member of the family, and the only girl, was herself involved in the war effort. She worked at the University of British Columbia, in a high security location, transcribing Japanese short-wave messages for the government. Norma had learned her skill at the side of her mother, Minnie Carson Bryson, who, I understand, was the first woman telegrapher in BC.

As an aside, my younger son Doug became interested in Morse Code when he was in Kindergarten and spent many hours at home with the encyclopedia, copying out the Morse Code for his own amusement. When he was in Grade One, he entered the School Science Fair at Ralph Bell Elementary School in Kamloops. I helped him build a simple telegraph machine. He was delighted to hear about the telegraphic ability of my grandmother, his great grandmother, Minnie, as well as Auntie Norma's war work. I am proud to say that Doug won a first place blue ribbon at the science fair.

After JB Bryson died, the brothers and their sister Norma decided to sell what had become the Bryson Co-Op Ranching Association under their stewardship to Colonel Victor Spencer of Spencer Stores in Vancouver. Colonel Spencer was amassing several ranches at the time. He first bought the Earls Court Ranch at Lytton, managed by Bert Erickson, where they raised pure bred bulls. Next he acquired the Circle S Ranch at Dog Creek, managed by Frank Armes, followed by the Carson Ranch at Pavilion. The Carson's Mountain portion of that ranch adjoined a portion of the Bryson Ranch on Pavilion Mountain. Finally, Spencer purchased our Grange Ranch at Pavilion; the combined ranches were later managed by my father, Clarence

Dad and Duffy were the only ones living and working on the ranch. The other three brothers and their sister, Norma Rositch, wanted some money out of the ranch to get on with building their own family homes. Since they were being pushed into selling the family-held operation, Dad and Duffy decided to put a ridiculously high price on the ranch in order to deter its sale. They decided on one hundred and fifty thousand dollars. When Colonel Spencer offered the family $100,000 for the entire going concern, Dad and Duffy were out-voted, although they thought that it was a very low price, even at that time.

We had to move on as a family and Dad ended up managing the combined Grange and Carson ranches, which Spencer called the Diamond S Ranch. This ranch was owned by Spencer and his partner, Frank Mackenzie Ross, later to be named as the Lieutenant Governor of British Columbia. The ranch used teams of horses for all the haying operations. Local Indians provided most of the teamsters, and, because it was wartime, about twenty Japanese would come from the internment camp in Lillooet to provide the labour: cocking hay and loading it on to team drawn sloops for ferrying the hay to the stack yards.

Later Dad moved on to manage the Chilko ranch near Riske Creek, in the Chilcotin. This was a huge ranch in the Chilcotin area of BC and it was a testament to Clarence that he was in demand as a ranch manager anywhere in BC. After we left the Chilko Ranch, Dad was offered a position to manage another big ranch—the Nicola Stock Farms near Merritt. Dad declined the offer from the owner, a Major Goldman, because the offer did not have enough money attached to it. Instead, in 1949, he used his share of the sale of the Grange Ranch to purchase the Voght Ranch, situated alongside the City of Merritt. The Voght ranch was owned at one time by one of Canada's first cattle ranchers. Voght had previously started the Douglas Lake Cattle Co., another of BC's premier ranches. Incidentally, I think BC is home to the five largest ranches in the British Commonwealth, Empire Valley being one of them.

Clarence inherited his size and strength from both of his parents. Minnie Carson was much younger than JB Bryson, who had been widowed early in life when his beloved wife, Mary Etta Currie, had died in her twenties in New Westminster. JB was left behind with their young daughter, Connie. After his wife's tragic death JB left Connie to be looked after by Mary Etta's parents. (In later years Connie would come to Pavilion to spend summers with her father and her half-siblings on the ranch.) JB had worked as a blacksmith in New Westminster after emigrating from Nova Scotia, along with three of his brothers, a sister and his widowed mother, in 1850. This was before the railway across Canada had been completed. Now he got the roaming bug again and moved to Ashcroft where he joined with a fellow named Smith to form the Smith and Bryson Blacksmith Shop servicing the horses and stagecoaches heading up the Cariboo Road.

Minnie was the eldest Carson girl. Like all the daughters of Robert Carson and Eliza Jane Magee, she was a big woman, commanding in stature. Ella, one of her sisters, was pronounced "Citizen of the Century" in Revelstoke. She had been hired there as a teacher by the school board in an effort to quell a revolution brewing in the local elementary school perpetrated by several unruly school children. Some of these "children" were in fact fifteen to eighteen years of age. That they were still in an elementary school likely had to do with the fact that they had to leave school for seasonal work each year. They were driving their poor young female teachers to distraction. Evidently Ella put up with no guff whatsoever and quickly whipped them into line. She was big enough to command the fear and respect of all of the young men, so she was a good hire for that school district.

The third Carson sister, Edna, was married to Detective Sergeant Morris of the Vancouver City Police Department. When I was going to high school at Britannia High School in East Vancouver I used to read the local Hastings News. In that paper I read that Edna had been charged and found guilty of assaulting her local butcher. Evidently she had accused him of putting his thumb on the meat scale while he was weighing her purchase. He objected vociferously, so she had "allegedly" reached across the counter, yanked him over, and bopped him in the nose.

The next time I read about her, she had split with Sergeant Morris and was driving taxi for a living. She had taken a fare to Stanley Park at night when the man sitting in the rear seat refused to pay her. Instead, he grabbed her around the neck from behind and demanded her money. She responded by grabbing his arms and yanking him into the front seat where she bopped him in similar fashion to the butcher. Again she was accused of assault by her nefarious fare. He had surely chosen to assault the wrong taxi driver.

Those Carson women all stood around five foot ten inches tall, weighed over two hundred pounds and they took no guff from anyone. I think my Dad and his brothers inherited some of that size and strength from Minnie as well as their father. JB Bryson was a big man who weighed in at about 250 pounds plus during his career as a blacksmith and farrier in the Smith and Bryson Blacksmith Shop in Ashcroft. Here he and his partner maintained the BX Stagecoaches on the Gold Rush Trail through the Cariboo. They forged shoes and shod the large horses that drew the stagecoaches and freight wagons along the trail.

**CaribooWagon Train—JB shod horses like these
Courtesy of Don Carson**

My old cowboy partner and family friend, Francis Haller, told me two stories about my grandfather, JB, whom he worked for as a young man. Francis recalled seeing JB working on the unruly and unbroken Clydesdale horses that arrived in Ashcroft after the railroad was completed in 1885. These big horses were destined to be teamed with previously trained horses that would pull eight and ten horse freight wagons up the Cariboo Road to Barkerville and the gold mines. JB's job was to make iron shoes using his coal fired forge to heat the metal. He was assisted by a hand pumped air bellows to fan the flames and a large metal anvil on which he pounded the hot metal. He would shape and attach these shoes to unbroken horses by balancing each of their bent knees on his own knee as he hammered each new shoe onto their hooves—most for the first time. As Francis recalled, when the big draft horses would struggle to avoid being shod, JB would expertly exert a lot of pressure on one hind leg, by pushing it up and back. This would lift the horse's whole rear end off the ground, getting the horse's complete attention and causing it to behave. This move was one way to bring about compliance that I, fairly adept at shoeing green broke horses myself, could scarcely imagine.

Since not many men were strong enough to handle these big horses, JB was known up and down the trail for his great strength. His reputation for strength was enhanced by the fact that as a young man he had injured his right wrist causing the wrist bones to fuse. Because of this wrist fusion he was able to carry the largest anvil in his shop with a straight arm outstretched. He could twist wrists with anyone and win, since his fused wrist was a secret weapon that kept anyone from being able to turn his wrist to the table.

Francis also recalled that a California roustabout had arrived at the Smithy one day, wanting to fight JB because of his reputation as the strongest man on the Cariboo Trail. The Californian challenged JB to fight but my grandfather ignored him.

"I've licked everybody up and down the Cariboo Road, from the Border to Barkerville. Now I intend to fight you."

JB did not respond to him at all, according to Francis, but he continued pounding and shaping shoes on his anvil. Nonplussed, the roustabout approached the large, heavy anvil to grab my grandfather. JB saw him coming and when the gent was within reach surprised him by grabbing his arm and twisting it until the roustabout was face first in the sooty dirt of the shop. Whereupon he jumped up and offered, "OK, JB, you're the better man!" With that utterance he continued back on his journey to California, presumably now the second best man on the Cariboo Road.

I witnessed an example of the strength that Clarence inherited from his father. Clarence was, at that time, managing the Chilko Ranch in the Chilcotin area of British Columbia. I had returned to live with the family after graduating from Britannia High School in Vancouver in 1949 and took my first paid job at the Chilko Ranch.

On a Sunday morning off, Dad and Tory Exshaw were testing each other's strength with feats to challenge one another on a beautiful summer day. They were twisting wrists and leg wrestling. My Dad would have been just under 45 years of age at that time, whereas Tory was a young, well-built, whipsaw-tough cowboy at the Chilko Ranch.

Dad amazed us all, and especially Tory, when he grasped a ninety pound anvil by the "horn" which is to say, the narrow end. He clasped his arm firmly against his side and stiffened his shoulders. He then swung the anvil horizontally out from the chopping block it had rested on and back to the block again. Tory couldn't do that, and I am doubtful if any man in BC could accomplish such a feat today. The combination of his good genes and a lifetime's worth of hard physical work had given Clarence's huge biceps and hands an awesome, almost mechanical strength. He and his brothers had apparently also trained to do this anvil stunt as young men, trying to keep up with their father. They learned to lift the smaller anvils this way, but apparently, could never lift the largest sized anvil the way their blacksmith father with the fused wrist could.

My Chilko Ranch job in the summer of 1949 was my first ranching job that was not done for room and board. My starting salary was a slim $2.00 per day. The first job I was given on the ranch was driving a team of derrick horses that pulled a sling load of hay to the top of a Pole Stacker. I was an old hand at this, having done that job on the Spencer Ranch for room and board. I found a new twist, though. I could ride the "double trees" that were attached to the team's harness and also attached to the cable that allowed the team to pull the hay to the top of the stack.

When one stack was finished, I rode the double trees behind my trotting team as we bounced over the fields to the next stacking site. This worked very well until one time I had the team on the run with me bouncing along behind when we hit a big bump which dumped me off. The team ran off at a gallop and when they got to a small cottonwood tree, one horse went to the left of the tree the other to right, and broke the harness all to hell. That was the end of that job.

Luckily George Mayfield's son Miles needed help on his cowboy crew and I was assigned to work with Barry "Red" Macue, Tornado "Tory" Exshaw and Maxine Mack. These experienced cowboys taught me everything I needed to know about the art of moving cattle.

Tory was the son of the famous rodeo cowboy, Cyclone Smith, who thrilled crowds at the Williams lake Stampede, where he often won the famous Mountain Race. He was killed in an accident on the rodeo grounds when he was acting as a pick-up man. Smith's horse collided with a furiously bucking horse and in the collision he was crushed underneath. After the death of his father, Tory was raised by his aunt and uncle, Hazel and "Ec" Exshaw. As an aside, Ec was the book-keeper for many years for Roderick Mackenzie at Mackenzies Ltd in Williams Lake.

I remember that first fall at the Chilko Ranch when we took a herd of five hundred head of the ranch's fat steers to the BC Cattlemen's Auction Yards in Williams Lake. Miles drove our chuck wagon, while we cowboys drove the herd for three days to the stock yards in "Willie's Puddle." Being the new guy, I was of course assigned to bring up the drag. The end of the drive coincided with a big celebration in Williams Lake, maybe held for Labour Day or Thanksgiving.

Chilko Ranch cattle drive to Williams Lake (photo Fred Waterhouse)

Miles used the occasion to get drunk at the big dance and the story we heard was that he made a pass at his cousin's girlfriend. For this he received a couple of punches in the nose which rendered him senseless. The next morning George Mayfield arrived at the chuck wagon where I had slept. Miles was still groggy.

"Harness up and get this 'blankety-blank' son of mine back to the ranch."

I harnessed up and we left on the trot with Miles asleep in the back of the chuck wagon. We made good time and arrived late that night at Bert Roberts' store and cabins in Riske Creek. Bert rented us a cabin and Miles and I went to sleep with no food at all that day. The next morning Miles came around and although he was nursing a big hangover, he cooked up some hotcakes for us. I was so hungry that I ate too many and couldn't look a hotcake in the face for a long time. As a side note—Bert's son Jimmy Roberts worked one summer for us as a cowboy at Empire, years later.

At the end of that fall on the Chilko, I went in to see George Mayfield about receiving my wages—my hard earned $2 a day.

"Unfortunately you have no money coming," he said. "You used it all up buying chocolate bars from the commissary."

Since I had only gotten in from the range a few times, I was totally chagrined. In those days this was a common tactic by big ranches, using commissaries as a way to avoid paying out wages. When I reported this to my father, he was very agitated and promptly quit his manager's job. So the cheap trick definitely backfired on George, for Clarence was as good a manager as he was ever going to get.

That was why Dad went looking for another manager's job in Merritt and why ultimately he used his portion of the sale of the Grange ranch to purchase the Voght ranch from a man named Wymond Pearl Sandy. WP raised North American champion purebred black Percheron horses. He had an auction sale of those horses when we bought his ranch. Those beautiful big Percheron horses brought WP an average of about $100 per head. At a similar auction in the US they would likely have averaged $1000 or more per head. We spent seven years at Voght Ranch in Merritt, from 1949-1956.

Dad followed in the footsteps of his father and his Carson uncles by becoming involved in politics. First he became a Merritt City Alderman. We did a a lot of volunteer work for the city as a family, turning some city-owned—but badly neglected—parkland into a beautiful park, complete with a running track. The city renovated an old grandstand and put in a new swimming pool; Voght Park is still operating today. Dad then decided to run in the 1952 provincial election as a Liberal candidate. That was the election when the Social Credit party burst onto the provincial scene. Dad ran in the provincial election hoping to win a seat and become the MLA for the constituency of Yale. The seat had been held for twenty-eight years by Dr. J.J. Gillis, until he stepped down in1952. Unfortunately, all Liberal candidates lost to W.A.C. Bennett and his brand new Social Credit party. That was the first and last time that preferential balloting was used in British Columbia.

Donald Gillis, the son of Dr. J.J. Gillis, married my sister Donna, in 1952 and later became a partner on our Empire Valley Ranch. Don was a tall, good-looking, well-built young man. He had dark curly hair, dark eyes, and a swarthy complexion.

Donna worked for the telephone company in Merritt as a switch board operator and so, as it turned out, she could keep track of who would talk to this handsome young doctor's son. I am sure she never stooped to actually interfering with a competitor's calls. Don managed his dad's Glen Walker ranch near

Merritt. Donna had become enamoured of Don after meeting him at a local dance. When they married in 1952, Donna was 21 years old, and Don was 25. Donna was pretty, fun-loving and easy-going. She ran for and was named the Merritt Rodeo Queen. She was a track star in Ashcroft and later at Britannia High in Vancouver.

After getting back from high school, where Donna and I had lived with my mother's sister and her family, my goal was to figure out how I was going to get to university. I wanted to attend UBC in Vancouver, but we couldn't afford the fees.

On the Merritt ranch we had only grazing range for one hundred head of cows so we supplemented our living expenses by raising and selling eggs from two hundred chickens. My mother sold eggs, milk and butter that she made from the milk that Dad and I produced by milking our three milk cows. Turnips was a Holstein while Snowball & Emily were Ayrshires. We raised five acres of potatoes and had a huge garden with a lot of raspberries for sale. We also did custom haying for many of the ranches in the Nicola Valley. We charged twenty-five cents a bale to cut, rake, bale, and stack the hay. Sometimes we made as much as $100 a day, which was a lot of money in those days.

It was a lot of work putting up hundreds of acres of hay and it was only made possible by my father's ability to keep an automatic hay baler working when other ranchers' balers had failed them. We were assisted by Albert Antoine, whose wife Bernadette cooked for us on the various ranches where we contracted to put up hay. I remember one day we all came in for lunch and Gordon—Albert and Bernadette's nine year-old—was playing with a sharp knife intended to cut the roast of beef that Bernadette had cooked. He cut his finger quite badly and blood spurted out all over him. Bernadette didn't hesitate;

she pulled the top off the kerosene lamp, poured the kerosene into a bowl and stuck his bleeding hand into the kerosene. This action immediately stopped the bleeding. This was a remedy I'd never heard of before. Gordon Antoine later became the Chief of the Coldwater Band, near Merritt.

Another incident from that time comes to mind. The Armstrong's Department Store butcher, Frank Haddad, had phoned to have us kill one of our veal calves which he would purchase to sell in their butcher shop. Since Dad was on the road politicking, it fell upon me to kill the animal. I asked Frank what he wanted me to do with the calf.

"Just cut its throat and hang it up," were his instructions.

I had never done this before on my own, so I followed that to the letter. When Frank came the next day to pick up the calf, it was not fit for human consumption, for I had not gutted it and taken out the entrails. I was distraught at my error but my mother suggested that it could be made into corned beef. I carefully cut out anything in the meat that was spoiled. From the edible portion of the carcass that was left after my trimming out the green stuff, my mother made marvellous corned beef and saved the day. We needed the money from the sale, which I had lost, but at least we had some good eating. I still love corned beef.

The last summer we lived on the Voght Ranch and did custom baling of other ranchers' hay, we felt that we had made enough money to buy a bigger ranch. Bob Carson Jr. was a cousin of Dad's and realtor out of Kamloops. When he came to see if Dad was interested in buying the Empire Valley Ranch that he had listed for sale from Henry Koster, Dad told him that we were. Through Bob, we ended up owning the Empire Valley Cattle Co.

Incidentally, we sold the Voght Ranch to friends, Ken and Doris Gardner and their three boys, Jerry, Jimmy and Ross, for $35,000, with the payments spread over five years. Shortly thereafter, the Craigmont Mine opened near Merritt and Merritt real estate boomed. Our Voght ranch became the southwest quarter of the City of Merritt. Within a year of the opening of the mine the hundred or so subdivision lots that were created out of the ranch were worth between $60,000 and $100,000 per lot. Would that we had held the Merritt ranch and sold it later ourselves. Ah well, *Woulda, shoulda, coulda*, as they say. (The Gardners did not profit much from the increased prices either because they sold the ranch before Craigmont Mines was developed). As for our family, the fact is that we needed the revenue we got for the Voght Ranch and used it to purchase our next ranch. In buying the Empire Valley Ranch, we entered a new era of large scale ranching and of wonderful backcountry adventures.

Back to Dad—He was a man of few words most times and he was not normally demonstrative. When he arrived home, he would hug the dogs and say hello to the children, but we knew were loved. He never spanked us if mother reported some misdemeanour, but his frowned look at us had more effect than a whipping. He doted on my sister Donna whom he called Lady Bird when she was a little girl. When people would drop by our home, we would hear my dad tell great stories and find out about what he had been doing when we were not around. One time dad arrived home from a rodeo, where he had been the announcer, with a black eye. When I quizzed him he claimed to have run into a door knob. Later I heard that he and his good friend Boyd, who owned the Flying U Dude Ranch at Green Lake, had gone to the rodeo dance. Evidently, Boyd had a little too much to drink and got into a fight. Dad, ever the peacemaker—and of course a teetotaller—stepped between the combatants and took the punch that was meant for his friend.

Dad was also very good at talking to me about things as we worked together and I learned a lot from him this way. Sometimes he was frustrated with me because I was so much smaller and slighter in stature than he was; like my sister Donna, I was small for my age. I did not grow to my adult height of six feet tall until after I graduated from high school. As my sister Donna liked to say, "I was called 'Tiny' and Mack was smaller I was!"

One time we were loading carrots and potatoes into our huge cellar. We had dug and sacked them into 60 pound bags. When we got to the cellar to unload them, Dad made no bones about admonishing me for not being able to hoist a "little" bag of carrots onto my back—he had one in each hand. That cellar could hold about four train carloads of vegetables and so our neighbour, Dick Steffens, would store his twenty-eight acres of harvested potatoes there. We grew five acres of spuds ourselves, which Dick would use his old Farmall tractor and equipment to plant and harvest for us. In return, each fall for two weeks, I would help him harvest his twenty-eight acres of spuds averaging twenty-two tons per acre. Every night, even though I was dog-tired, we would pack them into our cellar and Dick would spend his winters culling and bagging the spuds for local sales. For jobs like this, which I hated, Dad would always say, "Mack will do it." And, Bloody Hell—but I did it!

He used that phrase to my detriment another time. We had bought the latest automatic hay baler pulled by a tractor. However the "automatic" still required Dad to be seated at the rear of the bale chamber and I was seated on the other side. He would push two wires through a tripped block from his side, and I would tie the two wires around the bales before they would be pushed out of the chamber and fall to the ground. Each stroke of the plunger would cover me with hay dust and dirt. My face was black and my nose required blowing every couple of minutes. That was bad enough, but when we finished our hay fields Dad would lend the baler to other ranchers in the valley.

These other ranchers would hire a young lad to do my job. He would last a day, and then quit.

"No problem," Dad would say, "Mack will do it."

So that is how I spent my summer until a machinery company developed a real string tied "automatic" hay baler. The end result of all that dust in my lungs is that I now have pulmonary fibrosis. It may not kill me but I cannot climb any more mountains with my buddy Eric Paetkau.[6]

Dad was always in demand as an announcer at rodeos such as Williams Lake, Fountain, Lillooet and Minto City—before it was drowned by the Carpenter Lake Hydro Dam. He had a great voice and real presence in his western jacket and white cowboy hat. He had a habit of calling certain cowboys, "that *fameous* cowboy," with the emphasis on his added 'e'—that pronunciation drove me wild.

One time he overstepped himself and announced that an Alberta cowboy: "will receive nothing for that ride as he did not spur his horse out of the gate."

Of course, that was the job of the two judges, and so later the maligned cowboy and one of his friends tried to climb up onto the announcer's stand to teach Dad a lesson. Dad did a little jig on their fingers in time with the guy alongside him who was singing *My Home By the Fraser*. He said the jig worked and they climbed back down, literally and figuratively.

HOW MY DAD MET MY MOTHER

Eleanore Ferguson McCallum was the lovely younger daughter of Donald and Eleanore (Cowan) McCallum of Vancouver. They were from Ontario families that had gone west for the building of the Canadian National Railway and then moved south of the border to Minnesota with other railroaders. The town of Cowan, Manitoba is named for my maternal grandmother's family, or so I've been told. One branch moved to Vancouver. My mother, Eleanore, was a big-city girl with no experience related to a ranch wife's life. She had attended Britannia High School and Donna and I followed in her footsteps.

Coincidentally, when I first arrived at the Stevenson home in Williams Lake to take my future wife out on our first date, I learned that Liz's mother, Anne (Mackenzie) Stevenson, had also attended Britannia High School. Much to my surprise she told me that she had dated my Uncle Mac McCallum. Anne had been the captain of the grass hockey team and when she graduated from high school, my mother's older sister, my Aunt Elizabeth, took over the captaincy of that same grass hockey team.

As for my mother, she went to Sprott-Shaw College to learn to be a bookkeeper after her graduation from Britannia High School. This was a great deal of education for a woman in those days. Her family expected her to work as a bookkeeper for a short time and then to marry a well-heeled young man from the city, as a matter of course. This outcome was unexpectedly foiled when my mother decided to go to Ashcroft for the summer, after she finished her bookkeeping program. She and her friend Kay Pudney, as a summer adventure, went to work as canners in a tomato-canning factory in Ashcroft. She attended a dance, where my father spotted the delicately beautiful young woman with the raven hair and fair complexion, complemented by startlingly dark blue eyes that she inherited from her Scottish forbearers. Clarence was immediately smitten with "the girl in the red dress" as we heard about in his telling of the story over the years. Luckily for him, she fell for him just as hard and went home to tell her horrified family that she had met a fine young man named Clarence Bryson, whom she wanted to marry. She enjoyed telling us the story later of how she had described Clarence to her urbane father.

"He doesn't smoke and he doesn't drink and his name is Clarence," she proudly reported to her father, thinking of the McCallum family's Baptist heritage.

This apparently did not impress her father much. "He doesn't smoke and he doesn't drink and his name is Clarence—does he wear dresses?"

Her protective elder brothers, Bob and Mac, were similarly unimpressed. "If we had known you had wanted to get married, we would have introduced you to one of our friends."

"And if I had known that you wanted a ranch," her father added, "I'd have bought one for you."

Undaunted, she and Clarence set the date for January 5, 1931; they would be married in Vancouver. My father had little money to spare for anything, far less a ring and a wedding suit. The money luckily appeared when he got into a poker game with a haying crew and walked away with enough money to finance his part of the wedding—that was the one and only time he ever played poker for money.

Fortunately, he was not totally unfamiliar with Vancouver, for he had come down from the ranch at Pavilion to attend King Edward High School. He and his brothers took turns leaving their isolated family ranch at Pavilion to live with their paternal grandmother, Eliza Jane Carson, so that they could attend some schooling beyond elementary school, which was all that was available in their area. Eliza Jane had moved to Vancouver after the death of her husband, Robert Carson.

Clarence's main claim to fame at high school was being the best tackler on his King Edward rugby team but he was also a skilled mathematician. Education was very important to his mother Minnie (Carson) Bryson, who ensured that there were enough children on the ranch to keep a school open; she accomplished this by boarding other people's children to maintain the necessary quota of children needed to warrant a teacher for the Pavilion area. She also made butter and sold it in the Pavilion area to raise money for her children's education. Minnie inherited her appreciation for education from her mother's family.

Minnie's pioneer grandfather, Hugh Magee, was an irascible Irishman who was a well-known founding father of Vancouver. Magee High School in Vancouver is named after him. Hugh would row his eldest child, Eliza Jane, the fourteen miles from their homestead, in what was to become Kerrisdale, to New Westminster. She would walk home for the weekend if he was unable to come and get her.

MY MOTHER WAS A SAINT CALLED CAPPY

I'm not sure of the reason, but my poor mother was not well received by Minnie Carson Bryson, with whom they lived as newlyweds at the family homestead on the Grange Ranch. Perhaps she was viewed as being a soft city girl who was not experienced and efficient enough to be as helpful to the daily ranch operation as Minnie would have liked. I don't think Minnie was a whole lot different than many women of her time when it came to the appreciation of daughters-in-law.

When she arrived, my mother was expected to step in and start performing almost all of the household chores that were required, despite her lack of training in this area. My mother, a very sweet young woman, was quite defenceless to protect herself from her mother-in-law's harsh ways. Unfortunately, my father was not able to help her to deal with this unpleasant situation; these things happened indoors for the most part, and the men worked outdoors.

My father and his mother got along very well. Eventually Dad and Mother moved to a new home on the ranch which my grandfather, Donald McCallum, helped him to build. The new home was about a mile and a half from the main ranch house which had also served as a stopping house on the Cariboo Road. My mother often complained that her father and her husband had built the house across the road from the well. The house was quite distinct owing to a very tall peaked roof, which my grandfather McCallum was sure would be needed, for a flatter roof would surely fall victim to the heavy snowfalls in that part of the country.

My mother went to Vancouver to have her first two babies. My sister Donna was born at Grace Hospital. Fourteen months later I was born in my grandparent's house on 12th Avenue in Vancouver because my mother could not make it to the hospital in time. Duncan was born in Ashcroft, four years after my birth.

Despite her difficult relationship with Minnie, Eleanore learned to do everything required of her and do it well. She grew a huge vegetable garden, which necessitated canning and pickling. She learned how to use a cream separator so that she could make the butter that she traded to the First Nations community for buckskin gloves and slippers for our family. She became a great cook, doing all our baking as well as sewing and knitting. She knitted many beautiful sweaters and warm winter socks worn by everyone in the family and I still have some of her knitted Cowichan style sweaters and toques.

During the Second World War, she and fellow Women's Institute members of the community at Pavilion knitted socks and made quilts to send overseas to our troops. She learned to entertain the hunters as a hostess when the family began their game guiding operation at Empire Valley and also filled in as cook on that ranch when necessary.

Mother gave up a lot with the ranching lifestyle. She could not travel to see her family or friends in the Vancouver area very often and she was not able to attend church regularly when we lived too far from a community with a church, as was the case when we were at the Chilko Ranch, Empire Valley Ranch, and the Grange Ranch at Pavilion.

My mother's eldest sister Elizabeth had married a great guy named Bill Brett. When Donna and I finished elementary school, it was decided that we would go down to attend high school at my mother's alma mater, Britannia High School in Vancouver. We were to live with the Bretts in the family homestead on 12th Avenue, along with their six children. Donna shared a bedroom with her cousins Betty and Ruth and I bunked in with the boys, Billy-Don, Bob, Doug and Graeme. In a little house at the back of the yard lived my grandmother McCallum and her unmarried sister, Aunt Bess Cowan, who had come out from Minnesota. Donna remembers that they were very fond of each other but our grandmother was a night owl and Aunt Bess was a morning person. They struggled daily to accommodate one another in a small space so that the next generation could have the big house. They were both very wonderful, loving women who helped to care for all of us "unruly lot". We were all living under one roof and we all shared one bathroom.

My mother adored her Aunt Bess. She had gone to live with her in St. Vincent Minnesota, near St. Paul, for a year or so when she was younger. She loved that town and living with her aunt during that special time in her life. They were staunch Baptists, and my grandmother was one of the founding members of the Women's Christian Temperance Union in Vancouver. She had a great deal in common with my dad's grandmother Eliza Jane Magee and, because of this connection, they most likely knew each other. As a result of living in such close proximity, the Brett and Bryson cousins have always had a very close relationship, which continues to this day.

In the end, my mother fit into her new role as a ranch wife so well that you would not have known she was not born to it. As hard as she worked, she was always sweet and kind-natured and somehow retained some of her big city polish. Before she went to the Pavilion General Store she would dress up and put on her hat and her white gloves, as though she were going to town in Vancouver.

Mother adored her grandchildren and they were a very big part of her life. During our time in Merritt she saw Donna and Don's children a lot, for she and Donna had a close relationship. Then, when Don and Donna joined us at Empire Valley, they all lived together in the big house until a house could be built next door for them and their four lively and spirited children.

Empire Valley Ranch was a family run operation, with three generations living and working together. Monica was the first grandchild, and she was the apple of my parent's eye. She was always very outgoing and charming but when the younger children, John, Cheryl, and Kathy came along, Monica wanted to be reassured that she still held a special place in the heart of her grandmother, who was called Cappy by the grandchildren.

"You still love me best, right Cappy?"

"I love you all the same amount," Mom would wisely reply, "but I loved you first." That seemed to be the right way to answer the question and leave everyone happy.

My mother's bookkeeping skills from Sprott-Shaw College came in very handy on the ranches we lived on, but it was also a stressful role for her, for the money was always impossibly tight. My Dad would go out and spend money on needed equipment and things, leaving her to worry about how we were going to pay for them, while still repaying our operating cost loans and mortgages. She developed heart problems when we lived on the ranch and that was one of the things that we had to consider when we eventually decided to sell Empire Valley.

That decision was not taken lightly. We also took Dad's health problems (from bowel cancer) into account. We opted to sell the ranch in order to save Mother more stress and to allow her to live a more comfortable, and less physically and emotionally demanding life in retirement. The little ranch that they purchased at Monte Creek near Kamloops solved that problem beautifully. She was able to renew her church life at the old Anglican Church situated a mere five miles away at Monte Creek.

Mother passed away from a heart attack at age seventy-five. At the time, she was trying, with help from her granddaughter Kathy Gillis, to look after my dad. Kathy was attending Cariboo College (now Thompson Rivers University) and living with her Cappy and Grandad to help look after Dad. He was 225 pounds and had suffered a debilitating stroke at age seventy-five. They were inseparable all their married life and perhaps it was fitting that they were only finally separated by that which separates us all eventually.

A little aside regarding Dad's bowel cancer. In 1960, just before Liz and I were married, our family at Empire noticed Dad getting whiter and weaker. We insisted that he see a doctor in Kamloops. He reluctantly agreed. He claimed that he didn't like doctors but I think he knew that something was the matter with his body. He drove himself and Mother to Kamloops where a surgeon operated on him almost immediately and removed a large tumour from his bowel. Dad told the family that the tumour was benign. He returned to the ranch to resume life as though nothing had ever happened.

It wasn't until after we sold the ranch that I had occasion to find out more about what had been ailing Dad. I was hunting blue grouse on the hills above Knutsford (near Kamloops) with my friend Nick Kalyk and Dr. Clark from Kamloops. We had brought along a bag lunch and the doc and I sat down at an old cabin to eat while Nick was off chasing after his always-wandering dog, Ace. Dr. Clark, who was an anaesthesiologist, asked me about my dad's health.

"He's just fine" I said.

"That's amazing," he said, "because we took out a grapefruit-sized cancer from his bowel. He was supposed to come back regularly for treatments and examinations to see if the cancer had recurred."

This was the first any of the family had heard about Dad's cancer. That was pretty typical for Dad. To him, it simply, "wasn't that important."

On the lighter side, Clarence, at heart, was a big showboater who loved an audience. He particularly enjoyed playing tricks on the townspeople who came to the ranch at branding time. His favourite trick was calculated to catch the attention of all those seated and watching on the corral fence. Clarence did most of the castrating of the bull calves because he was a master at that job. So when Walter Grinder was seated on the ground holding the rear legs of a bull calf, Clarence would castrate the calf and pop the testicles into Walter's mouth. Walter would make like he was chewing and enjoying the feast. When most eyes were averted to avoid looking at this disgusting spectacle, Walter would quietly spit out the testicles on the side away from Clarence's audience.

Once Liz's cousin, Bruce Stevenson, came to our branding. He was wearing a white shirt and a pair of white jeans. Bruce watched for a while and then was encouraged to join in to wrestle calves in the corral to hold them while they were being branded. This was an open invitation for Clarence to wipe his bloody knife on Bruce's pant legs. At the end of the day Bruce's jeans were more red and black than they were white. (Bruce and his wife Pam own one of the biggest used bookstores in BC, in the city of Penticton, called The Book Shop.) Clarence did not, however, wipe his knife on the white jeans of his sister-in-law Dorothy Bryson, who would be inoculating calves against Black Leg and Malignant Edema. He was too much of a gentleman to pull that trick on Dorothy.

When our friends Mauri and Gwen Oaksmith and baby Ellyn came from Seattle to attend the branding, Mauri came to the cookhouse with me to have breakfast. Dad remarked on Mauri's nice clean khaki pants. Dad was in the process of putting peanut butter on his toast. "I'll just have to fix those pants for you," he said, as he proceeded to wipe peanut butter on Mauri's pants.

Rod and Anita Wright and baby son Kenny came from Vancouver to visit and they had a very long trip trying to find Empire Valley. I had told them to turn at the Dog Creek Store and then to go over Dog Creek Mountain. Unfortunately, the store had burned down and it took some wandering in the wilderness before the Wrights arrived at the ranch. The next day I was summoned to join the haying crew. Seeing my big strong visitor, Dad marshalled Rod, who was very much a city boy, into lifting bales. At lunchtime we went to eat in the cookhouse. Rod was amazed at the amount of food on the table and didn't think he could possibly eat such a big meal. That afternoon a crew came down to the fields bearing pies. "After all that work, I felt that I could eat half a pie by myself!" exclaimed Rod. We were grateful for the extra hands at a busy time.

My dad truly enjoyed life and was always ready for a good time. He worked hard when there was work to be done but his sense of humour and his love of fun were always nearby. He was a wonderful man, a true role model and I loved him very much.

Here is a portion of the eulogy for Clarence Bryson written by his grandson, (my son)

Douglas Bryson.

On July 22, 1908, on a small ranch at Pavilion BC, a son was born to John Bates Bryson and Minnie Carson Bryson. He was their third son and they called him Clarence George but his siblings all called him Dunc. He was my grandfather. We are gathered here to mourn his passing and to rejoice in his having lived.

He was a rancher all his life and he did what few of us in our lives can. He did what he loved, and he was the best at it. Throughout his eighty-two years he amassed a fortune; a fortune not counted in money, but in the vast numbers of people who loved and respected him. And he created a family of three children, nine grandchildren and nine great grandchildren whose names he remembered easily. There were, of course, boy and girl and boy—not to mention boy and boy and girl and girl. He lived his life in joy and optimism and at this time he would bid us not to be sad, but to peruse a while through these things he did which brought us happiness and laughter.

He is perhaps most famous for the proprietorship of the Empire Valley Ranch, one of the largest ranches in BC. There he ran cattle as well as a hunting and guiding business that was famous throughout BC and the western part of the United States. He ran Empire Valley from 1956 to 1967, during which time my brother and sister and I and most of my cousins were born. My grandad was famous also in later years for his pit barbequed beef roasts which he served at the end of a pitch fork. After Empire Valley, he bought a small ranch at Monte Creek, which he managed as well as helping out Joe Able and Ray and Pat Kerr manage their respective ranches. In 1984, he was struck down by a stroke which left him paralyzed on one side and unable to speak. That broke the hearts of anyone who knew him, to see him laid so low. He had been such a strong man and so full of life that he worked well into his seventies. But he was a fighter and he refused to give up. I remember the first words that he spoke after his stroke—upon seeing me decked out in my tuxedo for high school graduation in June, 1984. He said,

"Oh, Boy!"—I guess he forgot that the doctors said he wouldn't be able to speak. Despite his incapacitation, he still understood and recognized people and he retained his sense of humour.

I remember that he had the biggest hands and even years after his stroke he would envelop my hand in his and could easily crush it, despite my supposed size and strength. I remember as a child being afraid of him—he was so big and gruff and he always wanted to cut what he considered to be my long hair with sheep shears. But I grew to love his gentleness and his wry sense of humour. After his stroke, sometimes he would grab me and hold me by the hair, just to let me know that I wasn't getting too big to shear. When I went to UBC to play football I voluntarily shaved off all my hair and to see the look on his face when I came through the door—he was truly astonished! He kept on trying to grab my hair and just shook his head and said, "No!" In the years that followed, his health declined, but it never stopped him from enjoying his life and wanting to visit his friends.

As I said, we are gathered here in celebration of a life that has passed, and how lucky we all feel to have been a part of it.

[6] Eric and I climbed the Annapurna trail in Nepal and the Chilkoot Trail in the Yukon/Alaska. When we went to Peru to climb the Inca Trail, my son Jack came with us. Jack is married to Eric's daughter Carla, also a doctor. I think Jack wanted to ensure that both his children's grandfathers would come home alive. Eric has just published a book about his life as a country surgeon on the Sunshine Coast of BC, called "The Doc's Side" which is flying off the shelves of the book stores on the Sunshine Coast.

Chapter Eighteen
Range War—Friend or Enemy

To add to the Fred Froese saga, I recount two stories—After Fred had built his road up from Browns Lake through our Dry Farm Pasture and then onto crown land, he opened up a vast territory accessible by any and all in their vehicles. This was because the road now became a public road. Hunters had access to our privately held lands even though the land was fenced and posted with Private—No Hunting signs. They had merely to ignore the signs and climb through our fences in order to kill the mule deer that we continued to protect from wanton slaughter.

Fred had cut our barbed wire fences to build his road. He cut one fence at Browns Lake and one at the exit of his road from the Dry Farm Pasture. First, a story connected to the Dry Farm Pasture. Since we could not contain our cattle within the boundaries of the Dry Farm Pasture if there was a hole in the fence, I decided to build a cattle guard at the Dry Farm boundary. A cattle guard is simply a structure which cattle cannot negotiate. Typically, a buried cattle guard is built across a road that breaks a fence and takes the form of a grid of poles suspended over a trench. A vehicle can drive over a cattle guard but cattle cannot cross because of the spaces between the logs.

On a sunny day, Walter Grinder and I headed to the Dry Farm to build the cattle guard. I decided that Fred would just bring his D-8 Cat down to destroy a buried cattle guard, so Walter and I built a raised guard out of four inch jack pines spaced about a foot apart. When Walter and I had finished building the guard we proceeded back to the ranch buildings. On the way home we passed a point on the road where I could look across the Fraser River and see the road leading from Canoe Creek along the Fraser to the Gang Ranch Bridge, a distance of about twelve miles. If one crossed the bridge and made a left turn at the Gang Ranch turn off, one could then cross Churn Creek and drive up another twelve miles to Empire. On this day I happened to see Fred coming along the public road on the far side of the Fraser, in his bright red Land Rover. He was on his way to the new road that he had built, which began at Browns Lake. Realizing that it would take him the better part of an hour to travel the twenty-four miles, I drove back to the ranch and asked Dad and Dunc if they would drive up and protect the new cattle guard. Even though Fred, his hunters, or anyone else could easily drive across the new guard, I knew Fred's obstinacy about any impediments to anyone crossing onto the "public road"

Dad and Dunc drove to the new cattle guard, crossed it, and hid their pickup truck in the trees beyond. Then they took up positions behind big trees, quite close to the cattle guard. Soon Fred arrived to view our handiwork that filled in the gap of the cut barbed wire fence. He drove across the cattle guard and then backed his Land Rover up to the guard where he got out and pulled a chain out of his truck. He hooked one end of the chain on the cattle guard and the other onto his Land Rover, intending to pull the cattle guard apart. His actions, of course, were purely vindictive for the guard did not restrict access to the crown land.

However, when Fred straightened up from hooking the chain to his vehicle, he was startled to find Dad standing on one side of him and Dunc on the other. After a few shoves back and forth, Fred started to cry. Fred, through his tears, said that he would never touch anything of Empire's again if they would let him go. So they released him and he took off for his home like the mill-tails of hell were after him.

Of course, as soon as he got home, he got on his radio phone and phoned the RCMP in Clinton, and told them that he had been assaulted. A couple of hours later an RCMP Constable knocked on our door at Empire. My mother led him into our living room, where the constable accused Dad and Dunc of assaulting Fred. They denied knowing what he was talking about. The constable however, kept at it, berating the two men till my mother, who never normally raised her voice, told the constable, to "Leave and don't come back. If my husband and son say they didn't assault Fred, then they didn't." She ushered the Mountie out the door, little realizing that he had a valid argument. We heard no more from the RCMP—or Fred—until Christmas that year.

On Christmas Day, Don Gillis, Walter Grinder, and I had assembled about 200 cows from the Boyle Ranch at a narrow fenced-in area next to a cattle guard which separated the public road from our private property. This was also next to the point where Fred had received permission to build his road, as it was a school house lot. His road ran perpendicularly from our ranch road. We were cutting out those cows that we wanted to stay in the Boyle Ranch hay fields, from those that would be let through a gate alongside the cattle guard to go to the Bishop Ranch hayfields. We were well into our task when who should arrive but Fred and his son Jim. They backed up their Land Rover to our barbed wire gate. They pulled out pliers and were going to cut down the barbed wire gate till I rode up to them, mad as hell.

"Mack," Fred said, "you can't have a gate across this Public Road."

I got off my horse and approached him menacingly.

"Jim," he called to his son, "Get the camera, and get a picture of him hitting me."

Jim, doing as he was told, went to the Land Rover and came out with a Brownie Hawkeye camera. He stood in front of their vehicle and said, "OK, hit him."

I must admit, I was pretty furious by this time, although I had no intention of hitting Fred. Instead I grabbed the camera from Jim and smashed it between my hands. Then I grabbed Jim and threw the big ex-Marine up onto the front of the Land Rover.

"Now both of you get the hell out of here!!" I yelled.

They timidly departed, and we heard no more from them about the gate that they had intended to destroy on Christmas Day. Perhaps they had assumed we would be otherwise occupied; they somehow had missed out on the true meaning of Christmas.

After we sold the ranch to Bob Maytag, I heard that Fred had had a heart attack and died. I presume then that Fred's wife and Jim and his family packed up and went back to where they had come from in Washington State. They left behind several truckloads of stacked logs at Roaster Lakes that they had originally intended to use for building their log homes.

When I think of those home sites and the cabins that were never built, I remember Jim Froese asking me to pack him out by horse to survey their home site lots. (My time and the horses were donated, of course, just as the many meals we gladly shared with any and all of the Froeses.) I accommodated him and after a couple of days cutting out site lines that Jim was able to do by the aid of a compass, we planted the corner stakes for the lots. I saddled up the horses and turned them for the twenty-five mile return trip to Empire with the packhorse behind me.

"Wait a minute, you are going the wrong way," Jim said. "My compass points that way."

He wanted to go in the opposite direction from the one I had chosen.

"OK Jim, you go that way, and I'll go mine, or you can follow me, whatever you like."

I started for home and did not go very far before I looked back to see Jim following me on the horse that I had loaned him. I think maybe the rocks in that country had thrown Jim's compass off direction. I guess he decided that, in this case, the cowboy was more reliable than his compass. Of course, there was no way that an Empire Valley horse would have turned away from his buddies and gone in the opposite direction. The horse, in this case, was smarter than Jim, or at least, than Jim's compass.

Chapter Nineteen
Jerry Eppler— United States Marine

This big, raw-boned ex-Marine immigrated to Canada with Fred Froese because he was married to Fred's daughter Lucille. Jerry and Lucille weren't like the rest of the Froeses; they were gentle, friendly and sociable whereas the rest of that family would do likewise only if they had something to gain from doing so.

The Froeses located themselves at Porcupine Creek about twelve miles behind the Empire Valley headquarters, in a small frame cabin that had been trucked to that location by Arrow Transfer owner and CEO Jack Charles. It was to be used by Jack and his friends. Henry Koster had allowed Jack Charles egress to the site through Empire's privately held lands.

In addition to the small cabin, the Froese's had brought US army insulated tents for their family quarters. I guess Jerry and Lucille didn't fit into those tents because they were expecting their third child. They decided to move into the vacant Alfie Higginbottom hunting cabin about four miles closer to Empire after Alfie had finished his hunter guide operations in late November. The rat infested cabin had to be all cleaned up and winterized, which Jerry was able to do.

I think Lucille had seen a doctor in Ashcroft about their pending new arrival and they were encouraged to have the baby in the Ashcroft Hospital. However, Lucille decided to have the baby at home where her mother could help. So preparations were made to have the baby in the Higginbottom cabin, even though Lucille had almost died having her first two children in a hospital in Washington. Lucille's foster brother Dade and his wife Vi had recently had their baby in the Ashcroft hospital and then returned to the Porcupine army tents with the newborn. In the middle of winter, Lucile and Jerry had a premature baby boy at the Higginbottom cabin. As a matter of fact, it was on February 4th, my birthday. They kept the new-born baby warm by putting him into the warming oven of the old kitchen stove for the first few hours of the baby's life. To recognize the day he was born, Jerry and Lucille, named the baby Jerry Mack Eppler.

Jerry was full of stories about Marine life in Korea but one of those stories really caught my attention. It seems Jerry was home on furlough from Korea and had been putting the furlough to good use in the local bars of Washington State. He had been on a week's binge and arrived at his mother's home one night to find Mom waiting for him with jaundiced eye and critical comments. She scourged him about wasting his chances to visit with the family and relax from the stress of war. Mom had obviously been waiting for him for some time as it was very late. His tired mother was leaning against the outside wall of the walkway as Jerry approached. "Jerry—Drunk again!" exclaimed his mother disgustedly.

"Me too, Mom," said Jerry as he weaved his way indoors.

Jerry was obviously a big strong guy with large hands and big biceps. I got a kick out of him telling me about another time when he arrived home a bit *under the weather*. He discovered that his mother had locked him out of the house. Jerry grasped the door handle, which wouldn't budge. Jerry told me, with his American drawl, that he exerted extra pressure on the door knob and, "The damn thing came right off in my hand."

Not to make myself the equivalent of Jerry for strength, but this cowboy was not exactly a slouch in that department. One, time when I was in my thirties, my Barnhartvale Ranch partner Nick Kalyk and I had arranged with the Royal Bank in Kamloops to meet with the assistant manager in hopes of getting a bank loan to buy some cows for our little ranch. Nick, of course, was late—he usually was—and when he met me we got to the bank, not realizing that the four o'clock closing time had come and gone and that the staff had locked the door.

I strode up to the double doors but of course, they wouldn't open. I just thought they were stuck and so, like Jerry, I put a lot of extra pressure on the handles and the doors burst open to the accompaniment of horns blowing and whistles sounding. A bank employee, who recognized me, rushed up to tell us they were closed.

"We did not come all the way here to leave without meeting your assistant manager," I proclaimed, and lo and behold that gent appeared and welcomed us in—while the security staff went about fixing their broken door lock.

Chapter Twenty
CBC Tracks the Cowboys

Nobody was more surprised than yours truly when, in 1964, a team of CBC documentary film makers showed up at the Empire Valley Ranch. They explained that they were making another of CBC's famous documentaries, this one on the rivers of Canada. They were currently involved in a new river's contribution. This new movie would be entitled *The Fraser*. They were seeking contributions from a logger, a fisherman, a farmer, and a rancher. These were all to be situated along the Fraser River, which runs North from the high mountains West and South of Prince George to make a hairpin turn and then head South past Prince George then through Quesnel, Lillooet, Lytton, Boston Bar, Hope, New Westminster, and finally, approximately four hundred miles after its source, reaches the Pacific Ocean near Richmond BC.

The CBC crew consisted of a young blond female writer, a bearded male camera man and a bearded male producer. They took pictures around the ranch for a few days before they decided to follow me and our crew to Williams Lake for the annual Williams Lake Stampede, which took place at the end of June and took in Canada Day on July 1st. The CBC crew was to meet me in Williams Lake, but unfortunately a late run of snow from the mountains succeeded in washing out the Churn Creek Bridge behind them. This was our only way off the ranch by vehicle, and so a log footbridge needed to be constructed, as detailed in an earlier chapter. The CBC crew eventually hooked up with us and took pictures of us in Williams Lake, watching the Stampede and dancing in the Squaw Hall. The hall was an open air dance floor which was dangerous because disgruntled dancers, kicked out of the hall because of rowdiness and drunkenness, would throw empty beer bottles over the walls and onto the dancers. Squaw Hall no longer exists as it became dangerous, not to mention, politically incorrect.

After the Stampede ended, I spent a day rounding up our drunken crew to put them in the back of the ranch pickup truck. Each time I found one of our guys, I would buy a case of beer and put him in the back of the truck as I searched for the others. In this manner, after a few hours, we eventually arrived back at the Churn Creek log bridge to find the CBC crew waiting to video record the not yet sober crew crossing the log bridge. Fortunately, I think these pictures never made it into their movie. The way I got the men across was to have Walter Grinder, who was the soberest of the crew, lead across the logs with a drunken cowboy hanging on with both hands to Walter's shoulders. I followed these two and basically carried that cowboy across using both my hands on the back of his belt to lift and support him safely across. What a tandem sight that must have been!

After this filming was over, the CBC crew hung around for a few days while our crew assembled our next cattle drive to the mountain ranges. Then later, the arrangement was for the CBC to drive around by road to meet me at the Manitou Mine site which was at the confluence of the Tyax, Relay Rivers, and Mud Creek. Only one cabin remained in that area, of the many originally built there. The site was on our cattle route and was accessible by vehicles from Lillooet to Tyax Lake and then 10 miles up an old mine road alongside the Tyax River, to the Manitou mine.

On the appointed day, I led a string of saddle and pack horses to the CBC vehicle and hauled the crew up to our Bannock Cow Camp, situated on the Tyax Valley side-hill. This is where we had dispersed the current herd of cows, calves, and bulls into the alpine country that we referred to as the New Range. They took a lot of pictures as I introduced them to this incredibly beautiful BC alpine wilderness. After about three days of following me around and taking pictures, they determined they would like to be packed out. Since I was finished that cattle drive and would be heading back to the ranch to pick up cattle for another drive, they arranged to meet me in Lillooet.

They had brought my ranch pickup with them to the Manitou Mine site, so this allowed me to join them as they made their way to Lillooet. I was a few hours behind them, reaching the Lillooet beer parlour where we had arranged to meet. I was greeted by hoots of laughter when I arrived and when I questioned this rude reception they all stood up, pulled down their pants and chuckled, "I bet you thought you had worn the butts off us green horns, chasing us all over that wild country?"

It turned out that CBC had spent a few hundred dollars equipping them with specially made undergarments which prevented their legs from getting chafed. This explanation led to a good laugh and beers for all. We were all agreed that this was a good way to finish off a very hard few days of mountain trail riding. Of course, the best laugh was on me. Incidentally, I never did get to see the finished movie "The Fraser" and I really would like to see it, all these years later. So far, I cannot get it from the CBC archives.

Telling this story reminds me of something that happened several years after the CBC journalists had bared their butts to me. I am remembering a similar situation that occurred after I retired. I had phoned friends Jim and Shirley-Mae Jeffrey to see if they would like to come along on a seven day ride into the Chilcotin area, into the country where we had run our cattle. Jim informed me that they had just returned from a similar trip with the Menhinnicks. Liz and I had been invited to come for dinner to see the pictures they had taken on that trip where they had been accompanied by Brian and Audrey Williams. (Brian was later to become the Chief Justice of the Supreme Court of BC.) One of the Jeffrey pictures featured Brian and Jim showing off their legs encased in women's pantyhose. I thought they looked pretty silly—Jim in black and Brian in red—but evidently the pantyhose had prevented their legs from being chafed after spending hours in the saddle. I arranged our seven day ride shortly thereafter and told my buddies Eric Paetkau and Mauri Oaksmith about Brian and Jim's undergarments. We were in a Pemberton grocery store stocking up on necessities. They thought that the pantyhose sounded like a good idea so Eric asked one of the ladies in the grocery store if she knew where he could buy pantyhose and he explained why he wanted them. "Well, you'll need queen sized," she replied and directed us to a drug store several blocks away. When we arrived at the drug store it was to find a contingent of locals there to see the men buying the queen sized pantyhose. Eric and Mauri did not seem to be embarrassed, but I was. Our own companions, who included Bruce Hirtle, Joe Rideout, Nick Kalyk and son Jack, also got a big hoot out of the occasion. Not quite as big a hoot as later seeing Eric and Mauri in their finery getting ready for our ride into the mountains.

CBC MISSES THE COWBOYS BY A COUNTRY MILE

One fine fall day, we cowboys at the ranch got an unusual invitation. Could we come to Lillooet, our old home town, for a re-enactment of a scene from the famous "Hanging Judge," Judge Matthew Begbie's exploits as a Circuit Judge presiding over a vast territory stretching from the Lillooet Courthouse to the Prince George Courthouse. Begbie presided at the bench on all the serious crime trials for this huge area in the late 1800s. This was to be a part of Lillooet's Golden Jubilee Festivities or the BC Centenary, I am not sure which.

Dad, Don Gillis, Dunc, and I were to drive down to Lillooet where we would be met by our friend, Tom Christie, a rancher and game guide from the Mohaw area. Tom was to supply enough horses for us to form a posse and enough horses for Noel Baker and some fellow Lillooetans to form a phalanx of bad guys, reminiscent of the notorious Billy Miner Gang who in the past had robbed banks and railroads. The Miner gang was to rob the local bank and then repair to the old Lillooet Hotel bar where we, the posse, were to capture Noel after a short, furious gun battle. Then we were to proceed to practice frontier justice by tying his hands behind his back, throwing him aboard one of our horses and hanging him.

One of the objects of this endeavour was to have a CBC film crew capture the entire re-enactment on camera, including of course Billy (Noel Baker) being hanged on the famous Hanging Tree, in Lillooet. That tree had been made famous by Judge Begbie sentencing hardened criminals "to be hanged by the neck until they were dead."

After getting ourselves saddled up, our posse thundered down the main street, chasing worried tourists out of our way. We sprang off our horses and ran into the Lillooet Hotel bar to capture the "bank robbers." Dad had been told—he claims—that we could do anything we wanted while arresting the robbers. Dad roamed throughout the beer parlour kicking over tables of beer with his cowboy boots. A poor German tourist, unfortunately, did not understand the instructions given by the bartenders about our impending appearance. This petrified tourist climbed underneath an overturned table and pleaded for his life. We paid the German no attention. Instead we arrested "Billy" and proceeded to escort him on our way to the Hanging Tree. We roped his arms to his side and threw him on a spare horse.

Unfortunately, we got carried away and forgot that the CBC was supposed to be filming all this. They were a "day late and a dollar short" as the saying goes, for each of our stops. The final stop was the one where we were to hang Noel Baker, aka Billy Miner. Noel had contrived an elaborate leather vest with a hook at the top of his back to which we could attach a hangman's rope. This was all hidden under his jacket with the hook at the back of his neck.

We galloped up a fairly steep old gravel road to Lillooet's famous Hanging Tree, about three quarters of a mile from the beer parlour. I attached the already swinging hangman's rope to Noel's hook and slapped his horse's rump. This left Noel hanging and justice would appear to have been done. I guess we cut Noel, aka Billy Miner down, after a few hoots and hollers. I don't think we left him there, hanging and swaying in the breeze, but we may have done so. After the proceedings the posse members galloped back down the hill for a much needed free beer, for it was a hot day and we had been mightily exerted. Lo and behold, who should we meet half way down the hill but the CBC film crew coming up to film the hanging. My memory doesn't serve me well enough to know whether they were actually able to come up with anything for the CBC that would justify their being there.

After a beer at the Lillooet Hotel, Noel Baker invited me back to his home for a meal. It seems that he and friends had been hunting that spring in the Duffy Lake area near Lillooet, and had killed a young male grizzly during a legal season. The grizzly bear steak I was presented with was the most delicious and tender meat I have ever eaten. I can still taste it. This is said by a long time beef eating rancher.

Noel Baker, previous to our Hanging Tree trip had attempted, along with some friends, to run the Fraser River from Prince George to the Pacific Ocean in a little 16 foot boat propelled by a gas powered outboard motor. They had quite an eventful trip as they chose a time when the river was very high. Everything went along fairly well until they attempted the rapids at French Bar Canyon. Previous to tackling the dangerous rapids, they had beached the boat and pulled up the motor to check it over, fearing they might have damaged it earlier when they had hit a submerged log. When they dropped the motor back into the river they neglected to insert the pin that kept the motor in place. Their habit was to approach any rapids and then gun the motor to rush them through the rapids. At French Bar Canyon Rapids this system let them down and the unsecured motor flipped up. The boat also flipped over throwing all into the turbulent rapids. Noel says he went under till all he could see was blackness and he figured he was a goner. The next thing he knew the river threw him up on a bank a couple of hundred yards downstream. His buddies and the boat were all retrieved and they carried on minus all their personal items. I'm sure they never forgot to secure the motor again. Noel told me that the famous Hells Gate Rapids located further south on the Fraser, were a piece of cake after that mishap at French Bar Canyon. I can personally attest to that.

Chapter Twenty-One
Empire of Grass

GIANT DEAL LANDS RANCHLAND FOR BC IN BID TO PRESERVE GRASSLANDS shouted The Vancouver Sun headline of Feb. 10, 1998. Editor Stewart Bell called it a mammoth land deal, and it was. As an ex-owner of the Empire Valley Ranch, I applauded the NDP Government's decision, which was long overdue, but I wondered at the price and also wondered why the government of BC didn't buy it when it was offered for sale to them at a much lower price. The final cost of the deal was about 4.8 million dollars plus 1115 hectares of forested land along the Alaska Highway. (Evidently the forest land did not have to be cleaned up or reforested after logging).

"This will result," said Liberal MLA Richard Neufeld at the time, "in the loss of eleven million dollars in stumpage fees." This was a lot to pay, even for the purchase of one of BC's best known ranches.

The sale made me nostalgic for the good old days of cowboying on the Empire Ranch. Let's consider what was sold and at what cost.

According to reliable information, the Empire Valley Ranch could have been purchased for approximately 2.5 million in late 1994 or early 1995 from the estate of the late German national owner. The Environment Ministry would have had to have enough incentive to dig up that comparatively small amount of money. However, the politicians and bureaucrats dithered the deal away, probably thinking that they could get it for less if they waited; ranches in BC were not selling well at that point, and money in Government coffers was understandably in short supply in those days—of course it's even worse now. This left an opening for a couple of smart loggers from Prince George who moved in and bought the ranch at their price. They then moved in some of their logging equipment and—reputedly—sent a letter off to the BC Environment Ministry, threatening to log the ranch and make it look like a landscape on the moon if the government did not get off the pot and purchase it. The BC Environment Ministry quickly caved in and bought the ranch to save the grasslands for the California Big Horn Sheep that wintered on the ranch. To save this bunchgrass for the largest Big Horn Sheep herd in North America was certainly a laudable reason to buy the ranch but they could have saved the taxpayers a lot of money had they sent in a timber cruiser to find that one of the previous owners had already harvested all of the remaining timber that belonged to the ranch.

There was left only about 10,000 acres of forest owned by MacMillan Bloedel. That timber, remember, was part of the original deal which had seen the Brysons sell that tree farm in order to make a down payment on the ranch to Henry Koster. Empire Valley did not own that timber. It doesn't really matter. The government of the day had the first opportunity to buy the ranch cheap but instead they chose to dilly dally that opportunity away until they ended up buying it for almost twice as much.

Those Prince George loggers managed to *buy low and sell high,* whereas, our provincial government ended up *buying high, paying more, and saving some grass for the Big Horn Sheep.* The thing I don't really understand is that the grasslands were there when Mrs. Kenworthy, and later Henry Koster Senior, put the ranch together in its present form by buying out all of the small, Empire Valley pre-emption holding ranchers in the 1940s. That bunchgrass was still there when my family purchased the ranch from Henry Koster Junior in 1956. Make no mistake, after we sold the ranch, it was poorly managed and overgrazed by some of the successive owners over those thirty years between 1967 and 1998. However, the land and the potential for growing bunchgrass did not ever go away. Good stewardship would have, and will, restore the ranch to make it what it was when Henry Koster Jr. sold it to the Bryson family in 1956 and when we, in turn, sold it to Bob Maytag of Colorado in 1967. The degradation of the range started after Maytag sold the ranch.

The thing that may never come back, unfortunately, is the largest mule deer herd in North America which once wintered on the ranch. It was destroyed by a provincial government whose game management skills bordered on, to me, just plain stupidity. When we sold the ranch to Bob Maytag, because he was a US citizen, he did not want the adverse publicity that had stuck to the Bryson Family from incidents that arose while protecting the approximate 5,000 head of mule deer that wintered on the ranch. These deer used the grasslands along the Fraser River as a natural wintering ground, after funnelling in from the 1,800 square miles of wilderness back country which extended from Empire Valley to the Coast Range.

Harold Mitchell[7], a biologist and friend of mine, once claimed that: "Empire Valley was not an ideal wintering ground for mule deer and that there would be a big die off one day." Nevertheless, we had, in my estimation, managed the mule deer herd into a healthy vigorous population. We applied the same good husbandry that we used on our cattle herd to manage the mule deer herd.

We reckoned that to increase our cow herd we would have to ship only steers and dry cows and save the breeding stock. This we did, and our herd grew from 350 cows in the spring of 1957, to 1,500 in the fall of 1967, more than a 400% increase in ten years. At the same time, we ran a game guide operation on the 30,000 acres of land that we owned. The winter range for the deer flourished as did the bunch grass along the Fraser, by us expanding, with the help of the Forestry Range Department, our summer range out west towards the Coast Range.

We used the same principles for the deer that we used in building our cow herd. Our hunters were not allowed to shoot does, only bucks, so the deer herd had increased as dramatically as our cow herd had done. The bad publicity that Maytag had worried about came from our stewardship of the land and the deer. Our policy was to hunt over half the ranch with our paid hunters and to allow the use of the other half of the ranch by the public hunters. Because, generally speaking, the public hunters made no distinction between shooting bucks or does, they quickly cleared out the does on that half of the ranch while we cropped a surplus of bucks on our half. Any remaining does from the public hunting area smartly migrated to the side of the ranch where they were protected, proving that a deer is certainly not a stupid animal.

As time went by the bucks became very wary—except when they were rutting—and the does became almost domesticated under our system. This system worked very well for the deer and the cattle, but some of the public reasoned that they should be allowed to kill off all the deer on both halves of the ranch. Because of this reasoning, some meat hunters would frequently run headlong into my obstinate family who were not about to let that happen. We didn't own the deer but we did own the land, so we could pick and choose which hunters looked like sportsmen and which hunters looked as though they would do anything and break any laws they did not agree with, in order to get some meat.

Each autumn Dad's job was to separate the good legitimate hunters from the from the "undesirables" who would shoot up the country and kill almost anything resembling a deer. This he did very well with patience and fairness. Every fall he would sit out on the public road during daylight hours and advise well-meaning sportsman where they could hunt on our nominated public hunting lands and where they might have a good chance of bagging a deer. He would turn back those who often had a belly full of booze and those who were ready to shoot anything that moved. Of course, that led to a few misadventures and confrontations, but Dad usually had the last say in any arguments. We did have incidents and the odd scrap when Dad had to come back to headquarters for reinforcements.

Also, two of Dad's supposed *good hunters* set up camp where they were instructed, but after separating for a hunt, one hunter tragically killed his friend when he mistook him for a deer while they were working either side of one of the hundreds of timbered draws on the ranch. Despite those altercations, some bad publicity and a few court cases, we did save the herd, and when we left, the herd was at its peak. Bob Maytag, after he bought the ranch from the Bryson Family, turned the whole ranch over (not just half) to the BC Game Conservation Department, and they in their *wisdom*—and I use the word ironically—allowed all licensed hunters the right to shoot three deer, of which two could be antlerless. One didn't have to be a skilled animal husbandry person to understand what would happen next.

The first fall *hundreds* of does, their feet in the air, crossed the Churn Creek Bridge atop the vehicles of indiscriminate hunters—bucks were usually mounted in such a way as to show off their horns. What kind of stupidity allowed that kind of slaughter, and what kind of hunter could take pride in the killing of two tame does that had become used to ignoring real sportsmen intent on hunting bucks? Enough said.

On my last visit to the ranch, in the fall of 1996, I estimated there were approximately 150 mule deer roaming over those 30,000 acres of grasslands. These few were all that remained of the once proud herd of 5,000. So now the government aims to protect the largest California Big Horn Sheep herd in North America—it too has always existed on the ranch—and this is the real reason the government is purchasing these beautiful grasslands. Based on the NDP government's stewardship of the largest mule deer herd in North America, I am obviously worried about the California Big Horn Sheep. As the saying goes, "With those kinds of friends, who needs enemies?"

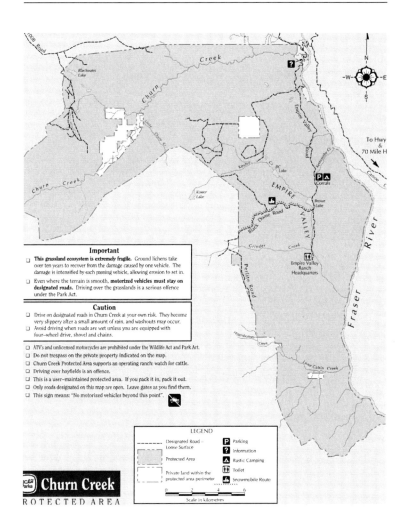

Churn Creek Protected Area—

[7] Harold Mitchell, unfortunately, lost his life in a helicopter crash while counting game. He was young, with a young family, so this was doubly sad.

Chapter Twenty-Two
French Bar Canyon Rapids

This story took place a long while after we had sold Empire Valley. A friend from Kamloops named Russ Chambers was well known there as a teacher, rugby coach and as a Cariboo College (now Thompson Rivers University) instructor where he was also the Women's Basketball Coach. I met him through my good friend Nick Kalyk. He had taught with Russ when Russ was teaching at Kamloops Secondary School. I met Russ during the summer when we were both helping Nick to tear down condemned buildings to recover bricks and copper piping. Nick was a junk collector par excellence. He later made a fortune by turning our farm tractor into a machine capable of travelling on train tracks to wind up all the copper transmission lines on the Canadian Pacific Railway after they had switched to microwave transmissions. They gave him the copper in return for winding up the lines.

In the fall of 1996, Russ called me in Vancouver to see if I wanted to go in his new jet boat up the Fraser River from the Big Bar Ferry to Churn Creek, on our old ranch. I jumped at the chance to see my old ranch from the river and invited along my son-in-law Paul Bucci, who is always up for an adventure. Russ wanted to start a new business transporting German tourists up the river to Churn Creek where they were to be met by a placer gold miner. Russ had previously arranged to allow these tourists to pan for gold on the miner's placer lease.

We went up the river twice on two different occasions. The first time we did the twenty mile trip it was a piece of cake going up through French Bar Canyon rapids. But on the second trip, coming back down through the rapids, it was quite a different story. The Fraser River narrows considerably as it almost makes a right turn at the French Bar Canyon. The standing waves generated by the river ploughing into the perpendicular rock walls are very scary for any boater, although it was reputed that a dry farmer from Big Bar Mountain had, many years before, navigated the rapids in a row boat, carrying a set of iron farming harrows.

Russ had to drive his jet boat like a car, sitting down behind his windshield with his foot on the accelerator. The big motor exerts a tremendous thrust by means of taking in water at the bow and having the big motor jet the water out of pipes at the stern. Our first trip down the river, I had stood alongside Russ holding on to the metal supports of the windshield. The fact that I was standing like this was fortunate, for when a standing wave would slosh water onto his windshield, Russ was effectively blinded. On that first trip I was able to direct Russ around the sharp bend in the river without incident, from my position standing above where the waves were hitting the windshield.

On October 27, 1996, Russ was finalizing preparations on his new enterprise and asked Paul and me to accompany him again. Paul and I followed Russ in his truck from Clinton while he towed his jet boat to Jesmond and then down the Big Bar Creek Road to the Big Bar Ferry over a very rough gravel road. The road from Clinton to Jesmond was covered with a fine layer of ice, so it was a bit tricky to traverse. Russ had brought along two friends from Kamloops that we met when we arrived at Clinton; Paul Sufferin operated a small café in North Kamloops, and his girlfriend Pat was a Swiss woman who was visiting from California where she worked as a tourist guide. Russ knew Paul Sufferin through the café for it was in the same building where he taught Aboriginal youth who were having trouble in the regular school system.

My son-in-law, Paul and I thought we would take advantage of Russ's invitation, take our rifles, and get off the jet boat at a place where we could walk up into a good hunting area of Empire Valley that we called the Fisheries. We had an effortless journey up the canyon where we could see herds of California Big Horn Sheep cavorting on either side of the river. Eagles were soaring over our heads and we were treated to an expansive vista that could only have only been exceeded by the famous Grand Canyon in Colorado.

We parted company with Russ and his two friends in order to hike up into the Fisheries for our hunting trip. We had arranged to meet Russ at our departure point at 3:30 that afternoon, which would give us enough time to get back through French Bar Canyon before dark. Paul and I had a great hike up into the Fisheries that brought back nostalgic memories but we saw only does— no bucks.

Fisheries Range at Empire. The Fraser river is unseen in the very bottom of the picture.

We had engaged in what Paul liked to call, "armed hiking."

At 3:30 we were back at the river and Russ was right on time. We climbed back into the boat with Russ and his two friends and proceeded back down the river. Paul and I knew enough about the Fraser to be afraid of it and so we were adequately protected. Paul had on neoprene fishing waders, belted at the waist, to keep water out should he unexpectedly end up in the river. Being even more wary than my son-in-law, I had two life jackets on, one on top of the other.

My respect for the power of the Fraser had been instilled at an early age when I had been on the banks of the Fraser and had seen a huge tree suddenly pop up from out of the murky water, seemingly out of nowhere. It had been held under water so effectively that there was no prior indication that it was there. I knew if a full grown tree could be held under water that long, it would be easy for a man to be submerged indefinitely. Paul

Sufferin and his girlfriend were not accustomed to this type of travel so it was a new type of adventure for them. As it turned out later, their life jackets were not good enough. We hit the French Bar Canyon rapids just half an hour before dark. The newcomers were seated up front to afford them a better view. Paul and I sat on the last of the three bench seats in the back.

As we headed into the rapids this second time, the standing waves again washed over the front end of the boat and obscured the windshield. That's when things quickly went wrong.

"Russ," I hollered at the top of my lungs, "you're heading straight for the rock wall."

He responded by cutting the boat sharply in a 180 degree turn and headed back up river. I am not sure why, but when he gunned the throttle, the front end of the jet boat dipped under the waves and we took on a lot of water. Since we were not making any headway up river, he again gunned the motor and the front end disappeared. We quickly submerged. I hollered at everyone to hang on to the boat because Russ had assured me that the boat's floatation made the boat unsinkable. Unfortunately, as it turned out, that was not the case.

Within seconds all but me had jumped ship. I was looking out the back end of the boat at the now rapidly approaching rock wall and when I turned around I was alone. Very quickly, the boat's prow dropped and most of the boat was submerged. I was left hanging onto a steel rear corner of the boat, while the prow appeared to be straight down below me, scraping on the rocks loudly enough for me to hear the terrifying noise. I jettisoned myself from the craft and quickly caught the others who were swimming to the rock wall.

"Don't go there," I hollered. "We can't get up that wall." (The perpendicular rock wall was about 150 feet straight up.)

They then turned down river and Paul and I got everyone to come together into a lifesaving circle. This was very necessary because with our arms around each other, we could buoy up Paul Sufferin and Pat whose life jackets were not very buoyant. However, a few hundred yards downstream we passed through a particularly large and turbulent whirlpool which broke apart our circle. I ended up clasping Pat to my chest to keep her afloat, while son-in-law Paul clasped Sufferin to his chest. Suddenly the wild whirlpool pulled Paul Sufferin out of Paul's grasp. Sufferin's life jacket did not stay fastened together on the front and he was not able to stay afloat on his own. He was quickly pulled under by the vortex of the whirlpool.

Pat began screaming, "Paul! Paul!" as she looked back over my shoulder to where Sufferin had disappeared.

My son-in-law, Paul, tried to swim upstream to reach Sufferin who was clearly drowning. Fortunately, the whirlpool picked Paul up and swung him on the upstream curve, since there was no way he could have swum against the powerful current. I looked around and saw that he was trying to save Sufferin. I thought that we would lose them both if he attempted the save.

"Paul," I hollered, "Leave him. Leave him!"

I did not want to have to tell my daughter, Lisa, who was at home with their baby son, Mackie, that I had lost her husband in the river. Luckily for Paul Sufferin, Paul ignored me, and the whirlpool carried him around the ring upriver so that in a one in a million chance, Paul's boot connected with Sufferin's shoulder, despite the fact that he could not see him through the murky water. Paul dived underwater and tried to thrust

Sufferin to the surface. He had to leave him and come up for air when they were both being pulled down into the increasingly cold depths of the river. After taking some big gulps of air, Paul attempted another rescue. He still could not see Sufferin, and Sufferin was even lower in the water, but somehow he found him.

For a second time Paul went down to try and bring Sufferin to the surface, all the while fighting the downdraft of the whirlpool pulling them both down. This time, despite all odds, he managed to push Sufferin to the surface. Paul thought that Sufferin was gone, and grasping him to keep him afloat he tried to shake him to consciousness as they were being swept down river.

"Breathe," he yelled. "Breathe you bastard, breathe."

Sufferin regained consciousness and started to scream.

We all continued down river in the cold rushing water with me clutching Pat and Paul clutching Sufferin, who was in shock. At that point I figured we would have to get everybody out of the river somehow, pull off our clothes, wring them out, and then run for the ferry, approximately three to five miles downriver. We were on the wrong side of the river for reaching the ferryman's shanty at Big Bar and because it was almost dark we would have had to holler loud enough that the ferryman's dogs would hear us and bark. My hope was that the ferryman would come out to see what the barking was about. Perhaps then he would see or hear us. That seemed to be our only chance in the desperate situation we found ourselves in.

Russ, meanwhile, had drifted ahead of us about fifty yards. Pat couldn't feel her fingers so I knew hypothermia was setting in. I attempted to swim us to shore at a bend on the west side of the river, but this did not work. The hydraulic action of the river kept washing me back. After being bounced back from the shore that I had intended to reach, we headed into a big bend in the river and Russ, who was ahead of us, started screaming for help.

I thought, *Jeeze, Russ, we are having a tough enough time without you suddenly losing it.* The reason for his hollering soon became apparent, though, as a jet boat like ours and a powered rubber raft were ashore on the ferryman's side of the river where their occupants had been panning for gold. They were packing up to go back to the ferry and from there to their vehicles, when they heard Russ's cries for help. We had seen that party in their jet boat earlier in the day, just below the rapids, and evidently they had wisely decided not to attempt going up through the Canyon rapids.

Russ's cries ended up saving our lives. A man came for Pat and me in the raft and I threw her aboard. Then I pulled myself in. The jet boat, crewed by two men and a woman, already had Russ and both Pauls aboard by the time the raft got there. I helped push Pat up onto the jet boat and followed by hooking my elbows over the three foot high gunwales, and then fell head first into the boat, exhausted.

As I clambered and fell into the boat I noticed blood spattering the deck below me and wondered where it had come from. One woman on the rescue boat turned out to have her first aid ticket. She was attending to the others up front behind the windshield and when she came back to the stern to help me, she took one look at me and her eyes bugged out. She grabbed a towel and jammed it onto a cut in my neck and then pulled my head over on top of the towel.

"Keep that pressure on your neck," she ordered.

Our new jet boat whistled down river to the ferry with me sitting in the howling cold wind where I developed hypothermia. The ferryman assisted us into his warm cabin and I asked if I could have a warm shower, since I was freezing. The first aid lady forbade this for she said I could have a heart attack from the sudden shock. So I undressed with Russ's help and he helped to warm me up by rubbing me down with a dry towel.

Looking back on that disaster, I guess none of us noticed that my neck was bleeding while in the river because my life jackets were pushed up above my cut, the cold river water slowed the bleeding and washed the blood away quickly. Our ferryman phoned for a helicopter to take us to the hospital. He was told that it was too dark for a helicopter but they would send two ambulances.

One ambulance was to come from 100 Mile House and the other from Clinton. The ambulances, of course, took a long time on those treacherous, icy roads. Paul Sufferin was still in a bad way and was in total shock. Pat rubbed him with towels, trying to warm him up. We had all warmed up by the time the two ambulances arrived to take us away. The Clinton ambulance prepared to take Paul Sufferin and Pat, whereas I was to go back to 100 Mile in the second ambulance. I protested that Paul Bucci would be taking my truck to Ashcroft and I could meet him at the hospital there. The medical attendant in the 100 Mile ambulance reluctantly allowed me to join the others in the Clinton ambulance after she had bandaged and examined me for about twenty minutes.

I found a cramped position behind the driver in the Clinton ambulance. Paul Sufferin was stretched out in the back with Pat and a medical attendant assisting him. Slowly, we maneuvered our way along the ice covered highway to Clinton where we had to be transferred to an ambulance from Ashcroft in case the Clinton ambulance received another emergency call. In Ashcroft my emergency doctor put eight stitches into my neck and bandaged my wound after telling me that the sharp steel corner of the boat had come within a millimetre of cutting my carotid artery, which would have killed me almost immediately. The gods were with me that day and I was incredibly lucky to have survived the ordeal.

As soon as Paul Sufferin had recovered, he and Pat were taken to Kamloops by Russ. My son-in-law and I stayed overnight with my brother Dunc and his wife Karyn in their home in Cache Creek, which is a short drive from Ashcroft. We had stayed with them on our way to rendezvous with Russ and the other two at the beginning of this nearly fatal trip.

In fact, the night before we left to meet Russ, Dunc had offered me his insulated mechanic's cover-alls, which sounded like a great idea, so I had agreed to take them. During the night, in a dream that seemed incredibly real, my mother came to me and said, "Mack, don't wear Dunc's cover-alls. You are going to be thrown into the river."

The next day, I refused Dunc's offer, which may have saved my life. That waterlogged suit would surely have pulled me under. That dream, on October 27th, came just two days after what would have been my mother's eighty-sixth birthday. Funny, I am editing this a day shy of what would have been her hundred and first birthday.

An epilogue to this story—we heard that our rescuers had gone up and down the river hoping to salvage our sunken craft, but to no avail; presumably it is jammed under a big rock and will be there for posterity. They also took the water temperature which was 32 degrees Fahrenheit. That was why there was ice at the shore, but not the river itself. The motion of the river kept it from freezing but it was freezing cold for us who were immersed in it. Our rescuers, from Sidney on Vancouver Island, have our everlasting thanks for saving us. Without that small chance of people being present to pick us up from the river, it is doubtful that any of us would have survived that frigid ordeal in the Fraser River.

I nominated Paul Bucci for The Royal Lifesaving Society medal and the Governor General's Award for Bravery. Both were awarded to Paul in 1998. He and Lisa were flown to Quebec City, along with Baby Aza (who was born after the accident) where Paul received his award from Romeo Le Blanc, Governor General of Canada at that time. (Nana and Grandad looked after two year old Mackie who was a going concern at the time, having inherited a lot of cowboy genes from some of his ancestors.)

This story was featured in Readers Digest and also a BC Adventure Series. A re-enactment movie was made featuring Paul Bucci and Bernie Fandrich of Kumsheen Rafting in Lytton BC.

Chapter Twenty-Three
Empire Valley Ranch— A Huge Cast of Characters

The Brysons purchased Empire Valley Ranch in 1956 and I arrived there in July, before the rest of the family, in order to harvest any available feed. The ranch came with its own cast of characters. First there were Louie Seymour and Harold Perkins, the teamsters who ran the horse mowers while I raked and baled the hay. Louie was an old one-eyed Indian in his seventies from Canoe Creek where he owned a small ranch; Harold Perkins was from Big Bar Mountain where his family had dry-farmed until the crops gave out. Then there was Slim Jepson, a chore boy who had blazed a trail from Empire to the Tyax River for Henry Koster; it was a trail we followed a few years later when we built the road from Yodel Camp to the Tyax. Slim was famous at the ranch for always blowing Taps on his trumpet in the early evening.

Walter Grinder and his brothers Henry and Johnny were also there when I first arrived. The Grinder brothers were responsible for getting Henry Koster's cattle herd back from the summer range so that they could be counted and sold to the Bryson family. They were from a small ranch along Big Bar Creek. The Grinder Creek place that was part of Empire Valley had belonged to a relative of theirs. Their mother was a First

Nations woman, and their dad was from Pennsylvania stock. Johnny Grinder left when the Kosters left in that fall of 1956.

Henry Grinder worked off and on for us as a cowboy and contractor. He finished building the Yodel Camp cabin that Alfie Higginbottom had started. Alfie Higginbottom did a lot of contracting for Koster and then did the same for us. Later Henry was a big help in building the pregnancy testing corrals on Gun Creek. His younger brothers, Walter and Art Grinder, worked for years with us as cowboys. Walter acted as cowboy, packer, and wrangler and was my right hand man. Art worked through that first hard winter of 1956-57, riding with me as I tried to locate and return to the hay fields our widely dispersed cow herd; the temperatures never got warmer than minus thirty degrees below zero for January of 1957. Hector Grinder came later to work as an irrigator.

Jimmy Joe from Pavilion, Dad's right hand man, and Harold Perkins from Big Bar Mountain worked year round. Then there was Henry Garrigan, a deaf guy from Big Bar who did a lot of fence contracting for us. As a young lad he had put Whites Liniment in his ears because he suffered from ear aches. In those days, with the lack of available medicines, folks would try anything; that particular cure rendered him deaf. Henry Garrigan brought with him four Belgian draft horses that he treated as family. He never worked those beautiful horses. When Garrigan and I discussed fence contracts he was always having to brush one of his horse's faces away from his so he could talk in his own particular way of signing. I became pretty adept at communicating through the use of hand signals.

Henry Koster stayed at the ranch with his wife Frances and their three young children until the cattle were all brought in and counted that first fall. The Koster family originally owned the ranch on both sides of the Fraser River. The sons split the ranch up with Henry taking the west bank, and his older brother Jack, taking the east side. Their sister Dodie was a partner in both her brothers' ranches. Jack died in his nineties, not that long ago, and his family still owns the BC Cattle Company at Canoe Creek.

After selling Empire Valley, Henry and family moved to Kamloops and got into the ranch real estate business. He also formed a partnership with Gordon and Ruth Abbey and my dad's cousin Bob Carson. They bought a ranch at Lac La Hache, which they named the Flying Five. The Abbeys had owned the Frontier Hotel—where we stayed if we overnighted in Clinton. The Abbeys became very good friends of ours. It was Gordon who introduced me to selling ranch real estate after we sold Empire. Gordon and I worked for Calvin Adcock at Adcock Realty in Kamloops but I gave up selling ranch real estate before I ever sold anything because I had an offer of another job. Calvin was also a partner of Gary Hook and Wayne Everett, both close friends of ours, in a ranch situated on the Trans-Canada Highway at Monte Creek. Henry Koster was very successful as a ranch real estate salesman and sold the Empire Valley Ranch a couple of times and the Gang Ranch a couple of times as well, I think. He and Frances could both fly their Piper Super Cub which was an advantage for him as a realtor, showing rural property. The couple broke up after they left the ranch and Frances moved to Penticton, where she ran a dude wrangling outfit. Henry died about 2003. His second wife Margaret is now living in Armstrong.

As I said, I gave up selling real estate without ever selling anything when Garth Bean, the sales manager of Rogers and Boyd Feeds of Abbotsford came into the Adcock Real Estate office and offered me a job as their Interior Sales Manager. I travelled from

Kamloops to Dawson Creek and back about every two weeks. I felt a bird in the hand was better than two in the bush, so gave up real estate. Liz and I had moved to Kamloops after the sale of Empire Valley. We had our good friends from UBC, Nick and Margaret Kalyk and Sandy and Ellen McCurrach living there. This made settling into a new routine much easier. Eventually I was promoted to run the southern Alberta operations of National Feeds, which had purchased Rogers and Boyd. I never took up the position because my next move would have been as the assistant to the general manager Stan Roberts, in Brandon Manitoba—even further away from the BC country which I loved. My UBC friend Ross Husdon had been offered that job before me but he declined also. Next after me, my friend Sandy McCurrach was offered the job, and he also declined the offer. We even had a going away party for Sandy and Ellen and gave them a clock to remember their Kamloops friends. The clock hung on their wall in their Kamloops house for years and became the focus of many funny stories. Stan Roberts wondered what it was about British Columbia that made these young men refuse to go east. He ended up coming to Burnaby, near Vancouver, bound for a job at Simon Fraser University.

Back to my family—When we bought Empire Valley, I moved up to hay what I could while the rest of my family finished the work on our the Voght ranch in the Nicola Valley, where we had lived for 7 years. The remainder of the family came to Empire in September of 1956. The Merritt ranch, as already mentioned, was sold to Ken Gardner and sons for $35,000 dollars, to be paid out over 5 years. In Merritt we ran about 100 head of cows up the Coldwater River, which is now bisected by the Coquihalla Highway. We eked out a living by raising 200 chickens and milking three dairy cows. We put in 29 acres of sprinkler irrigated grass pasture using Buckerfields' grass pasture seed. We summered 75 steers on that sprinkler irrigated pasture.

Buckerfields ran a competition to see what grass pasture in BC had achieved the highest production per acre using their pasture mix. We placed second to Alec Bulman of Westwold, and Dad's brother Duffy Bryson of Heffley Creek came in third. My mother, Eleanore, sold raspberries, eggs, milk, cream and butter. Her butter won the top prize for her every year at the Merritt Fall Fair. We also fed out 500 head of Nicola Stock Farms cattle through the horrendously cold winter of 1949.

That was how I met the famous Slim Dorin, who was the cow boss at the Nicola Stock Farms and earlier at the Douglas Lake Cattle Company. Slim and a few friends had started the Williams Lake Stampede in its early days. The winter of 1949, Slim and his cowboys drove the cattle down the unused railroad tracks the seven miles from Nicola Stock Farms to our Merritt ranch— the temperature was 40 degrees below zero.

I arrived at the Empire Valley Ranch in July of '56, and in September my dad, Clarence, my mother, Eleanore, and my brother Duncan arrived and we all moved into the main house at what we called the Home Ranch. Accompanying them was our long-time friend and worker Jimmy Joe. The ranch was situated on a bench above the Fraser River and the hay fields were scattered along the valley there. Empire Valley Ranch was composed of a string of smaller ranches that a woman called Mrs. Kenworthy had combined in the 1950s; these included the Boyle Ranch, Bishop Ranch, Grinder Creek Ranch, Magee Flats, the Bryson Pasture, the Brown ranch and the Zimmerley place.

My sister Donna arrived at Empire Valley in 1958 with her husband, Don Gillis—whom she'd married in 1951—and their four children—Monica, John, Cheryl, and Kathy. Don's father, Dr. J.J. Gillis, had just sold the Glen Walker Ranch that Don and Donna had been managing for him, twelve miles west of

Merritt. That Glen Walker ranch is on the Coldwater River along what is presently the Coquihalla Highway. Our old ranch ran 6 miles west of Merritt along the Coquihalla Highway to almost adjoin the Glen Walker Ranch. When the Gillis clan arrived at the ranch, their eldest, Monica, was around six, John was around four, Cheryl was around two, and Kathy was just a baby. Both Monica and John became part of the crew that worked as cowboys in the back country along with me and their dad. By the time they were in their early teens they were both good cowboys. Monica was my right hand cowgirl—I named Monica Mountain on the Tyax New Range for her—and John played the same role for his dad.

Monica on Tyax side-hill trail—with Don facing away

On the home ranch there was one main house that I shared with my parents and Dunc. There was also a large barn, a cook house, two bunk houses, and one smaller house that had been built for a foreman and his family, who chose not to take the job. The Kosters' plan had been to get more school kids along with the foreman, so that they could have enough children to operate a small school there. But when the foreman decided not to come, they decided to sell the ranch. Henry had been sent away to school as a boy; he hated that and didn't want to do that to his kids. This house that had been built and intended for that foreman later became the home for Liz and me and our young family. Liz and I spent time painting and renovating it. I put in wiring, plumbing and heating, with the help of my friend, Ray Emmerick from Merritt. Donna and Don lived for a time in the lower bunk house, which was pretty rustic. They built a new home when we could get enough funding to purchase a pre-fabricated home. Later, Duncan followed my lead and married his wife Karyn Fee, who was another young school teacher. Their marriage took place just prior to our leaving the ranch in 1967. Unlike Liz, Karyn wanted to be a rancher so she was mightily disappointed by the sale of the ranch. They lived in a house trailer on the ranch after they married. By the time we left the ranch in 1967 our family had grown to include fourteen members from the original four that moved there in 1956.

As I mentioned, Francis Haller had appeared, leading a couple of horses, shortly after my parents and Dunc arrived. We didn't know that he was coming but we were very happy to see Francis. He had worked for Dad on Pavilion Mountain when he was still married to his wife Leta. Francis and Leta had looked after the cattle for the Diamond S Ranch's owner, Colonel Spencer, when my dad was managing the ranch. Francis rode my dad's race horse Grey Ghost to win many mile races in the Lillooet area. Leta rode my mother's horse Princess to half mile victories. The Spencer ranch at Pavilion was formed by buying up the former Bryson and Carson ranches. Francis had always been a close friend of the family and he continued to be till the day he died. He was perhaps a few years older than my dad. He was a very athletic guy who could stand in one place and put his hands over his head and then kick his hand. He was a horseman, par excellence. He had a ram rod straight bearing, and he was a whiz with a rope and a horse. He was also an experienced bush-rider; I could not keep up to him following wild cattle in the bush. Francis was very good looking, and quite elegant, if you could say that without getting hit by a cowboy. He was also a bit of a ladies man. Francis was the grandson of the famous BC packer Cataline, a Basque from Spain, who was famous for his exploits as a packer on the Cariboo Road.

The Hallers were a well-known family in British Columbia. Francis had worked for my grandfather in the Smith and Bryson blacksmith shop that served the BX Stages on the Cariboo Road in Ashcroft when he was a young man. Speaking of stagecoaches—My wife Liz and her sisters, Rhona Armes and Jean Spence, along with cousins, Paul, Cam and Rod Mackenzie, donated the BX Stagecoach # 3, to the Williams Lake Stampede. This stagecoach had been purchased by their grandfather, Roderick Mackenzie. Roderick bought the stagecoach when he saw it sitting outside, exposed to the elements, in a farmer's field. He made an offer, the farmer accepted, and Roderick took the stagecoach to his home in Williams Lake and put it in a dry barn, which helped to preserve it. This stagecoach used to appear in the Williams Lake Stampede Parade every year, with Liz's mother, Anne Stevenson, often sitting on top in the luggage rack, dressed as a hurdy-gurdy girl with her front teeth blackened out. In 1958, during the BC Centennial, Princess Margaret came to visit Williams Lake during the Stampede. Doug Stevenson approached the planning committee to discuss the princess's visit and asked if they wanted to have her ride through town in the stagecoach. They declined the offer saying that the plan was that the princess would ride in an open convertible and wave to the crowd from there. On the morning of the parade the Lieutenant Governor's aide-de-camp arrived at the door of the Stevenson residence and told my father-in-law that Princess Margaret had heard about the stagecoach and wanted to ride in it. Doug was not pleased, as to put the stagecoach on display took a lot of time and work. A driver and a trained team of horses had to be organized to pull the coach. Luckily, local businessman and able driver, Claude Huston agreed to do the driving and horses were arranged. Doug attached the stagecoach to the back of his station wagon to tow it the three miles to town with Liz in the stagecoach. Amid the rocking and swaying of the coach on it's journey to town she was trying to get rid of the winter detritus of fir and pine cones left by the squirrels. As well, a Sunday School class had enjoyed a field trip to the Stevenson property and the remnants of their picnic needed to be removed before the the stagecoach was ready to receive the princess. New cushions were quickly picked up at Mackenzies Ltd., the family business located on the main street. (Now known as Mackenzie Ave.) The stagecoach then was ready for Princess Margaret. Liz's sister Jean (Stevenson) Spence was the Stampede Queen that year and was to follow the stagecoach on her horse. The Royal Canadian Mounted Police were in charge and Jean was told to keep up to the stagecoach. The horses pulling the stagecoach, having come straight off a rancher's field, were still in their winter mode and not quite ready to be pulling an unfamiliar "wagon". As the princess entered the stagecoach and rose to smooth out her lovely mauve summer dress, the horses bolted. This action threw the Princess back on her seat, and the horses proceeded to race through town. The townspeople, expecting to see the princess in a convertible, did not recognize that the young woman holding tightly to the side of the stagecoach was their Princess Margaret. Jean, mounted on her horse Sage, and following orders from the RCMP, raced behind the stagecoach in a fast gallop through town. When they reached the top of the hill to Stampede Grounds it was decided that the princess should get out of the stagecoach and go safely down the hill in the convertible. She was disappointed as she had loved her exciting ride. The next day the British papers were full of headlines such as: "Princess Has Wild Ride in Canadian Cow Town!" (Liz remembers playing on the stagecoach with her sisters and cousins when they were children. They loved to climb over the stagecoach to find the bullet holes made by bandits when the BX#3 was attacked while transporting gold south from Barkerville on the Cariboo Road.)

Back to the Hallers—when he arrived at Empire Valley, Francis came with his grey horse Spider and the humorously named Bartender, a quarter horse. Later he traded horses with Don Gillis. Don got Bartender and Francis got Whodat. Francis's son Dixie Haller came later, riding up river to our ranch from Watson Bar on the Fraser River, to the South of us, where he had a small place. He came to work for us, bringing two horses from a bucking horse string that he took to rodeos in the Clinton area. Buckles was a small buckskin coloured horse, and Salty a small white horse. They were what we'd call a cayuse, or a mixed breed of horses that were very tough; I think they may have been wild horses originally. They became two of my main horses when I bought them from Dixie. Buckles died of something we called Midnight Fever that he contracted when I penned him up at a woman's place near Tyax Lake when I went to attend Nick Kalyk's wedding. Salty would go anywhere with me and one time I led her down a twelve foot rock waterfall that I had got us boxed into and we had no other choice. We also got to know Rose and Irene Haller, Francis's granddaughters. Dixie contracted hauling and stacking baled hay in our earlier days on the ranch. Art Grinder worked for him on those contracts after he had left our Empire employment.

Jimmy Joe was a First Nations man from Pavilion who worked for my dad all his life and arrived at the ranch with Dad when they came from Merritt. He was around the same age as dad. They were both in their mid-forties when we bought Empire Valley. Jimmy's dad was named Joe Joe, and they were from the First Nations Reserve at Pavilion. Jimmy retired to 70 Mile after the ranch sold, and he died there. The missionaries who worked with the First Nations people could not pronounce the Native Indian names so they gave them all white men's names. They often doubled up on the same name for first and last names. So there was William Billy, Billy Louie (who took Donna and me to school, triple banking on a horse till we learned how to ride), Louie Louie, and Alec Louie. There was also Francis Edwards, the chief of the Pavilion Reserve, who served as our carpenter and his daughter Irene Edwards who cooked for us for years on the Pavilion Ranch.

We thought highly of Irene and referred to her as the "Princess." All these people worked for the family and were good friends. My mother would trade butter to the Ned family for buckskin gloves and slippers for our family. Not long after we arrived at Empire, Norman Ned arrived and he worked as a cowboy for several years. Sadly, some years after he left our employ, Norman was accused of manslaughter in Lillooet. His lawyer asked me for a testimonial which I provided. I remember Norman as a very quiet, taciturn fellow, who did his job well but seldom spoke. I was on a cattle drive in the mountains when his trial came and so I was not there to testify on his behalf and I have not been able to ascertain what happened at the trial.

Another early arrival at Empire was Eddie Narcisse from Fountain, who could run a Cat and became an irrigator with Jimmy Joe at the home ranch. Other cowboys worked more seasonally in the busy spring to fall. The cowboys and the field workers all lived in the bunk houses. Dixie Haller came to do hay contracting and brought his family for a few summers. Hector Grinder and his partner Jean Rocky and daughter Janice lived in the Bishop Ranch cabin and managed the hayfields there. Liz upset me once by giving my high school graduation suit to Hector who was not very big. It was the only suit I still owned until we were married. Of course, when I graduated from Britannia High School in Vancouver I was just 16 years old, 5 feet tall and weighed only 100 pounds so there was no possibility that I could ever wear it again. Later, Hector's family lived up behind our house in a tiny little irrigator's house.

Part time cowboys included Alfie 'none of your beeznus' Edwards and Marvin Tenault from Indian Meadows ("Let's get the show on the road"). There was also Norman Alphonse, from Alexis Creek, who broke horses for us as did Chris 'Cactus' Kind and Don McDonald; Cactus and Don had come over from the Gang Ranch. Amongst other cowboy jobs, Don had been a dude wrangler at Clinton, taking people on trail rides. He cowboyed on the Gang Ranch with Cactus. Cactus, whose name was actually Chris Kind, came to the Cariboo from Malta. He had worked on the Gang Ranch, where he changed sprinkler pipes in the hay fields. He had kept bugging them to become a cowboy because that's why he had come from Malta to the Canadian west. They eventually put him on a horse. He got bucked off into a cactus patch right away, so he became known as Cactus from then on. Norman Alphonse's sister Clara was Walter Grinder's wife and the first cowboy cook before I took over that job. She was also named Citizen of the Century for Alexis Creek. Liz and I met Walter and Clara's son much later, at the Williams Lake Stampede in July 2000, at the Millennium Celebration. We were in a line up and asked directions from a guy in line behind us. I asked him his name, and he turned out to be Walter's son, which was a great and wonderful surprise. We knew that Walter had children, but we had not known them when he worked at the ranch. They were living with their mother in Alexis Creek. Walter and Clara Grinder have both since passed away.

Alfie Edwards from Dog Creek was a real cowboy character. I would assign him tasks, and at the end of the day I'd ask him, "How'd you make out, did you find those cows I was talking about?"

"None of your beeznus!!" he'd say, which made all the other cowboys laugh.

I'd pressed him for more details: "What did you do today Alfie? Did you make it to the top of the mountain and find those cows?"

"None of your beeznus."

"But Alfie," I'd say, "it is my business because I'm your boss and I am paying you and feeding you."

Pressed again, he'd respond, "Dem cows way up down da udder side."

This sort of thing always got a big laugh from the other cowboys, and the joke became increasingly funny to them as the days went on. I kept Alfie around as much for the comic effect, as for his skills as a cowboy.

Kathy Hance was a young cook from Alexis Creek who worked for us one summer at Empire. The Hance brothers from Hanceville were genuine Chilcotin characters. When Donna and I arrived from high school in Vancouver, they were the first men we were introduced to by Wilf Hodgson of Hodgson's Truck and Stage Line after he had picked us up at the PGE Railway Station in Williams Lake. Dad would collect us from

Hanceville, which was only about a six mile drive from the Chilko Ranch. For ranch cooks we had Clara Paulson (Sam), ex-wife of Strawberry Sam a famous bronc rider from Pavilion. Clara was one of our longer-term cooks. Her then husband Jim Paulson worked as a chore boy when she cooked for us. Marion Eichhorn, from Kamloops, loved cooking for the hunters and the crew. She came back for years. Her son married my Uncle Mac McCallum's daughter Jeannie. Then there was, Maureen Lennox, another of our cooks who wrote wonderful letters to us later; and Helen whose last name I can't recall, but who used to sing at the top of her lungs: "Please Help Me I'm Falling, In Love Again" in a very unmusical voice.

Joe Matson was there for several years as a cowboy. He was young and tough, with an Alberta drawl. Joe had not had much education but he was extremely bright. He was always reading, and one of his favourite poets was Dylan Thomas. He also loved songs by Bob Dylan, a folk singer that none of us at the ranch had heard of at that time.

Helen Kerr and her husband Alvin Kerr worked for us helping during calving season for a few springs. Alvin's nephew, Raymond and his wife Pat owned the Harper Ranch on the South Thompson River near Kamloops. My Dad volunteered to help Pat and Raymond with the haying for several years after we left Empire. He was able to give them some good advice on ranching, using his wealth of experience. Raymond and Pat were able to show their gratitude to Dad by taking him and Mother on some wonderful trips that they otherwise would not have taken. They went to New Zealand, Hawaii, Texas, and Mexico with the Kerrs.

Erwin Dorsey was a part time cowboy for us who stayed on after his dad wanted to buy Empire when we had it listed for sale; sadly he could not come up with enough money. I think they later bought a smaller ranch in the Chilcotin. Bert Roberts from Riske Creek had a son named Jimmy Roberts who worked one summer as a cowboy. Dave Spillsbury worked one summer for us as a cowboy. I bought a big pinto gelding from him. The horse was big and tough and exactly what we needed in our back country but he resisted any attempts to be shod, which rendered him useless after a week on our hard, rocky trails. Dave's dad was the inventor and manufacturer of many of the marine instruments still used in boating on the BC coastal waters.

Epilogue
Life after Empire

We sold the ranch to Bob Maytag in the fall of 1967 and moved to Kamloops in 1968. Francis Haller moved to Kamloops with us and lived with Liz and me for a short time in our home on Columbia Street West. He then he moved in with my parents at their ranch at Monte Creek. He lived and worked as the head cowboy at the Harper Ranch for the Kerr family, where my dad worked as a volunteer, doing much of the haying operations, as well as building dams, irrigation ditches and laying heavy pipe from the main ranch to the riverfront hay fields. Francis was hospitalized for a heart condition and when he was released he lived at Ponderosa Lodge, a seniors' home in Kamloops, until 1971 when he had a final heart attack that killed him. Sadly, he didn't get to die in the mountains with his boots on, after all.

Walter Grinder went on to run his hunting camp at the back of Empire, hunting for deer and moose. He took on a partner, Frank Peter Meyers, a trained diesel mechanic from Germany, who had come to the ranch to work on machinery. Walter and Frank operated their hunting camp for a few years. Later, I heard that Walter had been tragically killed when he was struck by a vehicle in Williams Lake. I did not hear this news in time to attend the funeral, which I was very sorry about since I would have wanted to be there. He and I shared many adventures and were close friends after working together for so many years.

Joe Matson was a tough young cowboy. He was with us for a long time. He left the ranch when we sold to go driving underground trains at Bralorne mines. In Bralorne he married Marnie Dunbar, the Dunbars being a long established Bralorne family. They had one child, a boy named Mickey, after Joe's brother who had died young. Joe ended up ponying race horses at the Hastings Racetrack. He came to visit us once in Kamloops. He was a very bright guy who had come from a tough family life in Alberta. At the ranch he would bring Bob Dylan records up for Liz and me to listen to. At the time I did not know what to make of Bob Dylan but we soon came to know him as a major star when he was all over the radio. Depression resulted in Joe taking his own life.

When we left Empire, only two cowboys stayed on at the ranch to work for Bob Maytag. One was Cactus (aka.Chris Kind) and the other was Don MacDonald. Sometime after we left the ranch I saw Cactus in Kamloops; he came to ask me to borrow a mounted buck's head trophy to use in a Limestone Mountain lodge for a hunting and dude wrangling operation that he was helping to start. I heard that he had later guided hunters into the Pioneer Mine area, near Clinton, for California Big Horn Sheep—Pioneer Mine was the mine that Dad packed supplies to as a young man at Pavilion. I think Cactus probably did well there because it was prime sheep country. Later he got into riding bucking horses in the rodeos and he even made it to the Calgary Stampede. He wrote five books on the Cariboo. One was called *The Mighty Gang Ranch and Its Neighbor, the Magnificent Empire Valley Cattle Co.* He also wrote a book on bronc riding and another on sheep hunting. He has a profile at www.abc-bookworld.com. Growing up in Malta, Cactus had dreamed of being a cowboy, and he certainly realized that dream.

Our family moved to Kamloops after we left the ranch and my parents settled on a small ranch at Monte Creek, a half an hour's drive from Kamloops. Everyone worried that Dad would not be able to stand retirement and that it would likely kill him within a year but he found ranches to visit and to volunteer his services. He happily assisted Joe Able at Westwold for a couple of years and then many years assisting Pat and Raymond Kerr at their Harper Ranch. The Harper ranch had been started by Thaddeus Harper and his brother, who had started the Semlin Ranch at Cache Creek and also the huge Gang Ranch, our neighbouring ranch back at Empire. Dad revamped the Kerr's irrigation system by rebuilding their irrigation ditches and dam with a Cat as well as doing the haying on their big fancy tractor. My mother finally got her driver's license and traveled around from Monte Creek to visit her three children and their families. She joined a women's group called the Beresford Women's Institute and did volunteer projects with them. She was also able to attend church again which was an important part of her life that she had missed on our isolated ranches. She attended the historic Monte Creek Anglican Church. My niece Cheryl Gillis and her husband David Bathe were married in that tiny church. As an aside, when I was working as a citizenship judge I was to have a ceremony and a swearing in of new citizens in Kamloops. My mother had proudly invited all her Women's Institute friends to come to the ceremony. Sadly she had a sudden heart attack and died just weeks before the ceremony was held. Her friends came in her stead and we shed a few tears in her memory.

Barnhartvale Ranch—Mack, Liz, Doug and Jack hiding his face

Dad had a debilitating stroke in 1983. Mum would not hear of sending him to a home; the two of them were absolutely inseparable. She weighed all of about 120 pounds and Dad weighed twice that. Still, she looked after him, with the help of her granddaughter Kathy Gillis, until her heart attack in 1985. She was 75 years old. Dad then lived for about five years at the Overlander Extended Care Home in Kamloops before he too passed away.

My bother-in-law, Don Gillis, also died in the fall of 1985. When we left the ranch he had worked briefly as a partner in The Sand and Sage Hotel in Ashcroft. Later he worked in the mining industry and he helped to build the Revelstoke Dam. My sister Donna now lives in southern Alberta at Cardston, in a home beside her eldest daughter Monica and her husband Bob Wilson. Monica and Bob and their two children Randa and Riley compete in rodeos as well as train horses for rodeos. All four have been Canadian champions at their individual sport.

Bob, a retired teacher and vice-principal, cattle ranches and raises steers for use in the rodeo sport of steer wrestling, a sport at which he excelled. Monica was a Canadian Champion in the rodeo sport of barrel racing. She was selected Cowboy of The Year and her horse was Horse of the Year at the same time. Monica is the only woman ever to have won "Cowboy of the Year," a title that is voted on by cowboys. (Her grandad, Clarence and her dad Donald, would have been so proud.) She was also the only woman on the board of the Canadian Rodeo Cowboys Association; she has since retired from the board and daughter Randa has taken her place. Monica and Bob teach rodeo at Cardston Alberta, the only rodeo school in Canada.

John Gillis ranches in Alberta at Stavely. He has three sons, Clint, Ben and Rio. John was a champion competitive steer wrestler and now acts as a rodeo judge, as well. He was once a finalist for the "$50,000 winner takes all," Steer Wrestling finals at the Calgary Stampede. He looked as though he was going to win it all until the last of the five finalists decorated his steer .01 seconds faster than John. John inspects oil wells to augment his ranching income. Not many people can make a living out of just ranching without another "off the ranch" income.

Cheryl Gillis Bathe moved to Ontario where she is an esthetician. She and her husband David Bathe, a geologist working on oil wells in Alberta, live in a rural area near Minden Ontario, and they have raised two children, Adam and Eleanor. Kathy Gillis McLeod lives in Alberta where she works for the City of Edmonton as their manager, assessing property. She is married to Peter McLeod who is an actuary working in Abu Dhabi. They have two children, Donnie and Donna.

My brother Dunc and his wife Karyn had a ranch at Barriere for a short time and then moved up north to the Spirit River,

Alberta, east of Dawson Creek. They grew grain for many years while Dunc augmented their income by driving Cat, as well as laying out seismic lines for the Alberta government. Karyn continued to teach there. He and Karyn now live in the BC Interior at Cache Creek where Karyn was once teacher and vice- principal at the Cache Creek Elementary School while Dunc worked on neighbouring ranches at Ashcroft. Both are now retired and spend their time travelling in their motor home. They have two grown sons, Shane, who is in the computer business in Ontario, and Travis, who is an accountant in Nanaimo. Dunc and Karyn have four grandchildren, whom they adore.

As for me—Liz and I arrived in Kamloops in January, 1968. It was a time of exciting news in Liberal politics. Lester Pearson had retired, and there was a leadership convention coming up to elect a new prime minister. I won a spot as a delegate, as did my buddy Nick Kalyk, to go to Ottawa for the convention. At the end of the convention, Pierre Elliot Trudeau won the right to represent the country as prime minister. Almost at once we were into a federal election. Sandy McCurrach, Nick, and I were part of a delegation to try to convince Len Marchand, a fellow Aggie and a member of the Vernon Indian Band, to leave Ottawa, where he was working as an assistant to Art Laing, the Minister of Indian and Northern Affairs. Len was persuaded to come back to Kamloops, even though he felt that it might be difficult to defeat Davie Fulton, who had been a Progressive Conservative cabinet member as Minister of Justice in John Diefenbaker's government, as well as the MP for Kamloops and District for many years. However, with a tremendous campaign under the leadership of lawyer Jarl Whist, win we did, and we sent Len to Ottawa, to be part of the Trudeau Government. Len Marchand was the first Status Indian to be elected to the Canadian Parliament.

I did a number of different jobs after leaving Empire, including teaching school, selling grain, working as assistant to Len Marchand when he was a Minister of Small Business and later Minister of the Environment. Len was later appointed as a senator under Pierre Elliot Trudeau. He and his wife Donna, a former public health nurse, are retired and live in Kamloops.

I also ran for election as an MLA, in 1969, against Phil Gaglardi. Phil was the Social credit Minister of Highways who's most famous quote was "I'm the little guy that built the Yellowhead Highway." I didn't win, so I returned to UBC and completed a degree in Arts with a major in history and political science. I then went on and completed a teaching degree at Simon Fraser University. We returned to Kamloops to live full time and I taught Agriculture and Social Studies to Valleyview Junior Secondary School students and then later taught at an alternative school called McDonald Park. I took a leave of absence from teaching to run in the Kamloops Federal election of 1980 but lost to Nelson Riis. Nelson went on to serve 22 years as the Member of Parliament for Kamloops-Cariboo. I then returned to teaching.

Clinton Hotel—Old time politics—outside the old Clinton Hotel — courtesy Don Carson

Pierre Trudeau won that federal election in 1980. One of his last acts as prime minister before he retired in 1983 was to appoint me as a Citizenship Court Judge. When I was appointed I resigned my teaching position and moved to the Citizenship Court in Vancouver. I worked with two great judges, Norm Oreck and Madeline Basford (Nelson). I was based in Vancouver and travelled throughout the province to meet with immigrants who wanted to become Canadians. When my term expired, the government had changed and the Progressive Conservatives under Brian Mulroney came into power. As I served at the "pleasure of the government," my contract was not renewed. I took a financial planning course to become a mutual funds salesman, a job I was not happy with. My friend Ross Husdon then advised me that he was leaving his position as the manager of the Livestock Feed Board to manage the Okanagan Tree Fruit Authority. He felt that I should send in an application to replace him. I was interviewed, with several other candidates. I got the job and I ended my career working for Agriculture Canada. I was the Western Canadian Manager of the Livestock Feed Bureau (name change from Board) for nine years. When Brian Mulroney, the Prime Minister of Canada, chose to lay off 44,000 civil servants, he caught me in that net. I stayed on for a year with the Bureau, auditing our clients. I had barely left when I was offered a job as the Returning Officer for Vancouver East when the federal election was called. I was appointed by the Secretary of State, and in that position, I managed four federal elections in Vancouver East.

Liz went back to teaching when we moved to Kamloops. She taught Grade One at Dallas Elementary School for twelve years and when I was offered the job at the Livestock Feed Board, Liz moved to Vancouver and continued to teach Grade One at Georges Vanier Elementary School, in Surrey, where she taught for another twelve years. When she retired, some of the teachers she had mentored came to speak of how much Liz had done to help them in their careers. Although she was sixty-five years old, Liz was not anxious to retire but Jack and his family were going to travel to New Zealand for a year and we wanted to join them for a couple of months, so retire she did. Though she has never looked back, she does miss her days with her young students.

Lisa continued her early interests in plants and animals. When I was taking my teaching practicum in Kamloops, Lisa became interested in entering the Science Fair to be held at Ralph Bell Elementary School, using her aquarium to do her project on The Learning Process of Fish. She won the exhibit, then entered the Kamloops Science Fair and won the right to enter the BC Science Fair in Vancouver. She was in Grade Six, and at the end of the Fair she had won the Biology Trophy for the whole Science Fair, from Grades Six to Twelve. She also won the overall prize for all areas, from Grades Six to Nine. We were all overwhelmed, for she had had almost no help from us. The spelling on her write-up was uncertain, and her backdrop was somewhat wrinkled, having come down in the back of our station wagon from Kamloops. But she knew her stuff and could answer all the questions put to her, with aplomb and accuracy. Upon graduation from the University of Victoria, where she majored in psychology, Lisa became a social worker. She married her high school sweetheart, Paul Bucci, and they have two children, Malcolm Douglas Mackenzie Bryson-Bucci (Mackie) and Aza Mari Elisabeth Bryson-Bucci, as well as Paul's two children from his first marriage, Andrea and Paul-Alex Bucci. Paul and Lisa live in Vancouver, where Lisa is a social worker and Paul is Director of Digital Products for Pacific Newspaper Group.

Jack had a successful athletic career at Kamloops Secondary School, as well as getting top marks academically. He was on the

AAA basketball team that went to the BC finals. He joined the KSS football team that won the BC Championship in 1981. On graduating from KSS, Jack went to UBC and became a family doctor. He married Carla Paetkau, whom he met at medical school, and they both practice on the Sunshine Coast, where Carla grew up and where they are involved with sports and community activities. They have three children, Malcolm Eric Bryson, Duncan Mark Bryson, and Ingrid Elizabeth Bryson. They are close enough to Vancouver that we are able to have frequent visits back and forth.

Doug, our youngest, is the only one who does not remember living at Empire Valley, for he was less than two years old when we left. He was also on the football team at KSS and, because he never wanted his brother to get one up on him, his team also won the BC High School Football Championship. Doug was honoured to win the Frank Gnup Cup for Defensive Player of the Game. Doug went to UBC on a football scholarship and his team, the UBC Thunderbirds, won the Canadian Championship Vanier Cup in 1986.

Liz and I attended all the boys' games, driving from Kamloops to Vancouver for an evening game when we had to return that night for two or three hours of sleep before arriving at work in the morning. We also went to Toronto for the Vanier Cup games. Doug went to Willamette University College of Law in Salem Oregon and practiced law for a while in Portland. There he met and married Marcia Harrison, a woman from North Carolina. They did some travelling and then Doug decided that he loved the law but he didn't enjoy being a lawyer. Doug and Marcia now live in the Kootenays, in Nelson, BC, where Doug can get in as much time in the mountains back-country skiing, biking, and white-water kayaking as he can. He is a Cat Skiing guide with Big Red Cats, out of Rossland, and an avid volunteer with Nelson Search and Rescue. Marcia works as a social worker in Nelson, working with developmentally disabled adults. They have bought some rental duplexes which they have worked hard to renovate. Their two dogs are their pride and joy and, as of this final edit, they are all getting ready to take a year off and drive to the Southern Tip of Argentina.

Now Liz and I are permanently retired and we could not be happier because we live in Vancouver, close to our three children and five grandchildren. All our spare time is now spent travelling and visiting friends. I hiked some of the major hiking trails of the world with my buddy, Eric Paetkau, our son Jack's father-in-law. Liz says Jack and Carla's wedding was a "marriage made in heaven" for the two fathers who met and became fast friends. We love fishing, hunting and hiking. We go every year as guests of Jim and Brenda Rough to their fishing lodge called Black Gold on Rivers Inlet. We hiked the Annapurna Trail in Nepal, where we were known as "The Cowboy and the Surgeon" for the help Eric gave to injured hikers, assisted when necessary by the cowboy, acting as nurse. We also hiked the Inca Trail in Peru, a trip which included Jack and Jimmy Oaksmith, as well. Carla says Jack had to go along to keep the two grandfathers out of trouble. We also hiked the Chilkoot Trail in the Yukon and Alaska, and we canoed down the Yukon, again with Jack and our friend Mauri Oaksmith as well as Eric's son David who is a movie and television actor. (Currently in the police drama—Flashpoint)

Eric, Dennis Rogers and I hiked the Rainbow Mountains' Trail in BC with my brother-in-law, Gordon Armes who supplied the horses, ably assisted by a mutual friend, Hoberly Hove. On the Rainbow Mountain hike I took a short-cut from one trail to the next in what turned out to be the worst jungle in BC. I camped overnight in the bush next to a big fire. The next morning I made my way south to where we had left our trucks. Unbeknownst to me, my buddies felt I may have had a heart attack, so Gordon rode in and used my vehicle to drive late at night to Anahim Lake to alert the RCMP. The next morning two helicopters and 20 volunteer search and rescue guys from Anahim Lake were ferried out to look for me. I was confronted by a helicopter and four Mounties on my arrival at the trucks. Talk about embarrassed, to say nothing of the cost to BC taxpayers. People say I was lost but I call it "misplaced." Joe Rideout from the Livestock Feed Bureau did that trip as well and goes with me whenever horses are involved. (He ran the Maritime section of the Livestock Feed Bureau while I ran Western Canada. Joe and his partner Lorna took Liz and me on a tour of Newfoundland where we saw wonderful country and we even met the Premier of Newfoundland, Danny Williams.)

Many years after we left Empire, Liz and I rented Barry Menhinnick's horses to take Jack and Carla's families and Lisa and Paul's families to Spruce Lake. Malcolm, Duncan and Mackie were very young. They loved the hiking, riding and fishing at Spruce Lake. Of course they all would like to go again, especially Aza and Ingrid who were not yet born when the boys went on the mountain adventure. On an earlier occasion Donna Gillis, Lisa and I rode into Spruce Lake. Jack was still at medical school and Carla was doing her internship at Lion's Gate Hospital. They decided to take the long weekend out of their very busy schedule to ride in on their bikes to meet us there. The weather did not cooperate and their bike wheels soon became mired in mud and they had to carry them. Carla told Jack that she would rather be in labour than carry that heavy bike covered in mud through the rain and up the muddy trail. We, in our tents at Spruce Lake, decided that we should go to meet them as they were very late in arriving. Luckily the horse riders arrived in time to save the day. Auntie Donna had a hot stew and a warm tent to welcome the frozen, exhausted cyclists. In the morning Carla woke up, pulled back the flap on the tent, saw that it had snowed in the night and uttered, "I can't believe it but Hell has just frozen over!"

I also made two seven day horse rides with Barry and Warren Menhinnick to the back country behind Empire. Jack, Mauri and Jimmy Oaksmith went on those rides as well as Joe Rideout, Nick Kalyk, Eric, and good friend Bruce Hirtle.

We often travel with our friends from UBC. Ellen and Sandy McCurrach organized a trip to China for ten of us, as well as accompanying Liz and me to Cuba. Elliot and Alison Sclater planned a riverboat cruise from Amsterdam to Budapest for many of our UBC contingent and from there we went on to Prague. We had so much fun that we decided to go to Ireland together on a Rick Steves Tour the following year. We went to France with Sandy and Ellen on another Rick Steves trip. We decided to go to Normandy first to visit the Canadian War Memorial at Juno Beach before joining the group for a tour of Paris and Southern France. After the tour we got on a train and went to the Cinque Terre in Italy where we hiked the five villages. Several years later we joined Ross and Doreen Husdon and went to the Adriatic on still another Rick Steves Tour. After the tour ended, Ross became our guide. We went to Sarajevo and took a side trip to Visegrad to see the Bridge on the Drina River. Ross had us read a book by the same name in preparation for the journey. For our 50th anniversaries we went on cruise through the Panama with Sandy and Ellen McCurrach and Dean and Wendy McLean.

My cousins that I grew up with, Bill Brett, Betty Brett Webber, and Bob Brett are no longer with us, and Ruth's husband, Bob Armstrong—my friend from Merritt days, is now gone. My much loved cousin, Helen Bryson Rowan left us much too soon. My Aunt Norma Bryson Rositch and her son (my cousin) Rob Rositch passed away about the same time. Many years before, Helen's brother Gary died a hero when he flew his crippled T-33 jet trainer into the ground rather than parachute to safety. The pilotless plane would have crashed into the people in the town he was flying over. My good buddy Nick Kalyk died of a heart attack in 2006, while playing tennis in Kamloops. We also lost our dear friends Elliot and Alison Sclater. Our cherished Kamloops friends, Ernie and Pat Klapstock, Rick Walley, Bimal Verma and Jarl Whist were all lost much too early. A lot of friendships made and some, regrettably, gone.

So many people come into your life, some playing big roles, others playing only minor parts—so many names that it is hard to keep track of them all. I met a lifetime's worth of folks in the eleven years spent at the Empire Valley Ranch. As time has moved on, I've learned that life is like a train that travels in but one direction and you never know when you'll be getting off. If there is something that you have always wanted to do, my advice to you—my motto—is <u>do it now</u>!

About the Author

Mack Bryson grew up on family ranches and worked on the ranches from the age of 12. He graduated from Britannia High School in Vancouver in 1949 at the age of 16 and then took his first paying job for $2.00 a day as a cowboy on the huge Chilco Ranch. Next his family bought the Voght Ranch in Merritt which was sold seven years later in 1956 in order to buy the much bigger Empire Valley Cattle Company, West of the Fraser River and next door to the Gang Ranch. His adventures on the Empire Ranch form the basis of A Cowboy's Life. The ranch was sold in 1967 to Bob Maytag of Colorado.

Mack's first job after the ranch was in Kamloops, as Interior Manager of National Feeds. He was promoted to manage Southern Alberta in 1969 but decided that he didn't want to leave BC and so he resigned to run as a Liberal Candidate in the BC provincial election, under Liberal leader Pat McGeer. He lost to Highways Minister Phil Gaglardi. Mack then returned to UBC and obtained a BA, majoring in in history and political science. He went on to Simon Fraser University where he earned a teaching degree. He taught high school in Kamloops before becoming Minister's Assistant to Len Marchand, Minister of Small Business and later Minister of the Environment. After Len retired in 1981, Mack ran in the Federal Constituency of Kamloops Cariboo, but lost to Nelson Riis. In 1983 Mack was appointed by Pierre Trudeau to the Citizenship Court of British Columbia to act as a Citizenship Court Judge where he served for three years. In 1987 he became Western Canadian Manager of Agriculture Canada's Livestock Feed Bureau, for ten years. After retiring from Agriculture Canada in 1997 he finished off his career, serving as Returning Officer for Elections Canada through four federal elections in the Constituency of Vancouver East.

Mack and Liz have been living in Vancouver since 1988. They have their three children, their spouses and five grandchildren living nearby.

Grandparents—Malcolm, Duncan, Liz holds Ingrid, Mackie wearing Granddad's cowboy hat, Mack holds Aza, circa 2000. (Normand Photo)

Acknowledgements

Many friends and relatives have helped in the writing of this book. I am afraid I am going to leave some of you out, so in advance, I'd like to say thank you to all. You've helped to make this book a reality and my life a fun and exciting adventure.

First, I would like to thank my cousin Rita Bryson Morrison for the hours and hours of research that she put in to compile a book on our family history. Her book enabled me to gain information quickly and easily. And thank you to my cousin Don Carson for his knowledge of family history and old pictures which he shared with me.

Thank you to my children, daughter Lisa Bryson, who helped shape the beginning of the book, helped with typing of the manuscript and was always there to offer historical and literary advice and encouragement. Thanks to her husband Paul Bucci who helped me to survive in the French Bar Rapids. A vote of thanks to my son Jack Bryson and his wife Carla Paetkau, who helped to keep my body as healthy as possible and my mind in one piece. I am grateful to my son Doug Bryson for his legal and literary advice and to his wife Marcia Harrison for her wonderful "Marciaritas," which helped me to recuperate after a long day at the computer.

Thank you to my grandchildren: Malcolm Bryson, for his expert typing, Duncan Bryson, for computer expertise when his grandad was mired in ignorance, Ingrid Bryson for her love and encouragement, Mackie Bryson-Bucci for typing and for helping me with a recalcitrant computer and to Aza Bryson-Bucci for her accurate typing and her ability to make her grandad smile.

Thank you to my buddy Eric Paetkau for always setting the goal post high and for keeping his notebook up to date on "Dumb Mack Incidents."

Thank you to Ellyn Shull, author and family friend, whose advice on the title of the book and help on the synopsis were invaluable. Also, to her parents, our long-time and loyal friends Mauri and Gwen Oaksmith, whose advice and suggestions were very much appreciated. Voyages in the brisk sea air are a valued part of our lives.

Thank you to the Honourable Len Marchand and his wife Donna, our dear friends, for keeping the dates organized and political events in order and to Len for his invaluable knowledge on the grasslands of BC. To our buddies and faithful friends Sandy and Ellen McCurrach for their help and advice, and for adding encouragement when I flagged. And thanks for the memories of many wonderful times together.

To Linda Ewart, for permission to use her father Peter Ewart's painting of cattle crossing the Gang Ranch Bridge, on the back cover of the book. The timing of this picture leads me to believe that the cowboy in the picture is me on our "First and Last Beef Drive." The hills in the background show a portion of Empire Valley. This painting was on the cover of the BC Telephone Book in 1959.

Thank you to my Britannia High School buddy Bob Donnelly and his son Alan, who found old ranch pictures, printed and delivered them to me—also for remembering stories from the "old days."

Thanks to my cousin Lynne Bryson for sharing her memories of her mother, Norma Bryson Rositch's time working as a wartime telegrapher at UBC. Thanks to my cousin Glenis Bryson for her help with the early history of Lillooet. Her mother, English war bride Hilda Bryson, was in charge of the Lillooet Museum for many years. Susan Bryson Bell is now at the helm of that museum. Thanks also to my cousin Doug Brett for finding and delivering a book on Empire Valley when I needed it at a crucial time.

Thanks to sister-in-law Jean (Stevenson) Spence for her memories of Princess Margaret's visit to Williams Lake and to brother-in-law Gordon Armes for all he did to try to recover me when I was "missing in action" in the Rainbow Mountains.

I want to thank Margaret Kalyk for her memories of "Old Whitey's" life after Empire.

Thanks to Myrna and Jim Evel for always giving us "a place to land" when we came to Ontario, and for our on-going, special friendship.

A special thanks to Darci, John and Harry Swinton who came to my aid by delving into the interior of the computer to find missing information. They came just in time to prevent a nervous breakdown.

My love and thanks to my sister Donna Bryson Gillis and my brother Duncan Bryson, who shared these adventures with me and who were always "along for the ride."

My love, gratitude and thanks to my wife Elizabeth for sharing our life together. She was, she is and she always will be the love of my life. Liz' devotion to this book has sometimes exceeded my own. Thank you Liz!